Introduction to Estimating for Construction

Students and professionals encountering estimating for the first time need an approachable introduction to its principles and techniques, which is up to date with current practice. This textbook explains both the traditional techniques and best practice in early contractor involvement situations, within the framework of modern construction procurement.

As well as introducing different estimating techniques, it includes:

- the nature of costs in construction from a cost of resources approach;
- modern tendering procedures, and the stages of development of construction projects;
- how to convert an estimate into a formal tender and then into a contract;
- simple numerical examples of estimates;
- estimating and cost analysis during the construction project;
- summaries and discussion questions in every chapter.

This is an easy-to-read introduction to building estimating for undergraduate students, or anyone working in a quantity surveying or construction commercial management role who needs a quick reference.

Brian Greenhalgh FRICS FCIOB FQSi is a contract manager currently engaged by a major client organisation in the MENA region. He was formerly a principal lecturer in quantity surveying and construction project management at Liverpool John Moores University with responsibility for postgraduate programmes in quantity surveying and construction project management.

Introduction to Estimating for Construction

Brian Greenhalgh

LONDON AND NEW YORK

First published 2013
by Routledge
2 Park Square, Milton Park, Abingdon, Oxon, OX14 4RN

Simultaneously published in the USA and Canada
by Routledge
711 Third Avenue, New York, NY 10017

Routledge is an imprint of the Taylor & Francis Group, an informa business

British Library Cataloguing in Publication Data
A catalogue record for this book is available from the British Library

Library of Congress Cataloging-in-Publication Data
Greenhalgh, Brian.
Introduction to estimating for construction / Brian Greenhalgh.
p. cm.
Includes bibliographical references and index.
1. Building—Estimates. I. Title.
TH435.G665 2013
692'.5—dc23 2012021438

ISBN13: 978-0-415-50986-2 (hbk)
ISBN13: 978-0-415-50987-9 (pbk)
ISBN13: 978-0-203-08006-1 (ebk)

Typeset in Sabon
by Keystroke, Station Road, Codsall, Wolverhampton

Printed and bound in Great Britain by the MPG Books Group

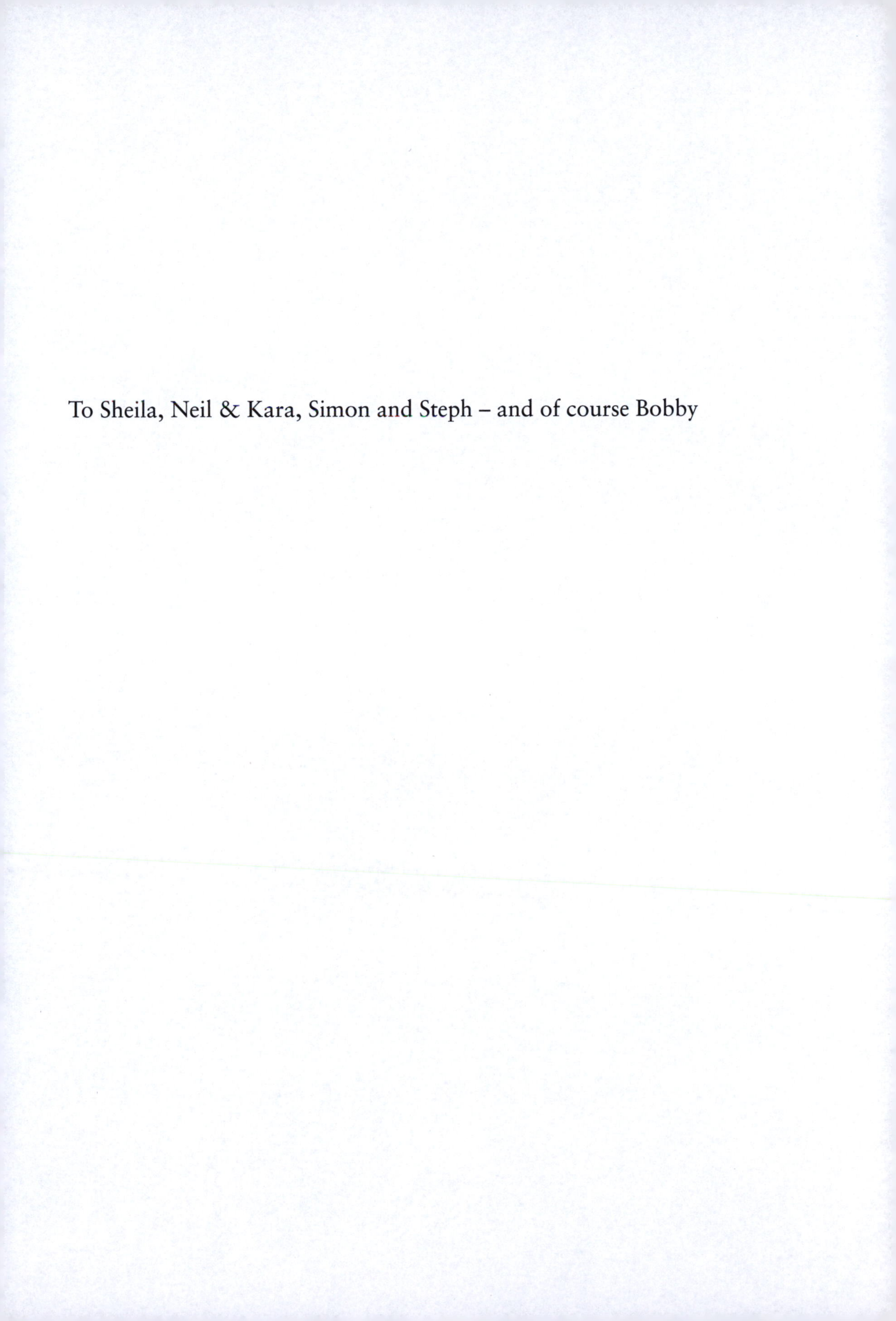

To Sheila, Neil & Kara, Simon and Steph – and of course Bobby

Contents

List of figures

List of tables

List of abbreviations

BAFO	Best and Final Offer
BCE	baseline cost estimate
BCIS	Building Cost Information Service
BIM	Building Information Modeling
BLS	baseline schedule
BOQ	bill of quantities
BPS	baseline project schedule
CAR	Contractor's All Risks (Insurance)
CAWS	Common Arrangement of Work Sections
CDM	Construction (Design and Management) Regulations
CESMM(4)	Civil Engineering Standard Method of Measurement (fourth edition)
CIJC	Construction Industry Joint Council
CIOB	Chartered Institute of Building
CIS	Construction Industry Scheme
CITB	Construction Industry Training Board
COSHH	Control of Substances Hazardous to Health
CSI	Construction Specifications Institute
CV	Curriculum Vitae
CVR	cost value reconciliation
ECI	early contractor involvement
EMC	estimated maximum cost
EPC	engineer, procure, construct
EPCM	engineer, procure, construction management
EPIC	engineer, procure, install, commission
EUR	element unit rate
EVA	earned value analysis
EVM	earned value management
FEED	front-end engineering and design
FFE	fittings, furniture and equipment
FIDIC	International Federation of Consulting Engineers
GFA	gross floor area
GIFA	gross internal floor area
HR	human resources
HSE	health, safety and environment, *or* Health and Safety Executive

HVAC	heating, ventilation and air conditioning
ICC	Infrastructure Conditions of Contract
ICE	Institution of Civil Engineers
ICV	in-country value
IRS	information release schedule
ISO	International Standards Organisation
IT	information technology
ITT	invitation to tender
JCT	Joint Contracts Tribunal
JIT	just in time
JV	joint venture
JV-MC	joint venture main contractor
KISS	keep it simple sweetheart
KPI	key performance indicator
LCC	life cycle costing
MCD	main contractor's discount
MEAT	most economically advantageous tender
MENA	Middle East & North Africa
MEP	mechanical, electrical and plumbing (installations)
MGP	maximum guaranteed price
MOU	Memorandum of Understanding
NEC	New Engineering Contract
NJCC	National Joint Consultative Committee
NRM	New Rules of Measurement
NWR	National Working Rule
OFT	Office of Fair Trading
OGC	Office of Government Commerce
OJEU	Official Journal of the European Union
PC	prime cost
PCSA	Pre-Construction Services Agreement
PFI	private finance initiative
POMI	Principles of Measurement International
PPC	Project Partnering Contract
PPE	personal protective equipment
PSC	Professional Services Contract
PTE	pre-tender estimate
QA	quality assurance
QM	quality management
QS	quantity surveyor
RFI	request for information
RFP	request for proposal
RIBA	Royal Institute of British Architects
RICS	Royal Institution of Chartered Surveyors
SCM	supply chain management
SMM(7)	Standard Method of Measurement (seventh edition)
SWMP	site waste management plan
TPCS	total project cost summary
VAT	Value Added Tax

VE	value engineering
WBS	work breakdown structure
WLC	whole life costing
WRA	Working Rule Agreement

Foreword

In January 2011, the sister book *Introduction to Building Procurement* was published by Routledge, which I co-authored with Dr Graham Squires of the University of the West of England. As Danny Myers quite correctly pointed out in the Foreword to that book, it had gone through a long period of gestation, but the final product was a notable contribution from someone in the twilight of his career! I take this as a compliment as it is everybody's duty at this stage in their career to pass on any acquired knowledge, learning and, dare I say, wisdom.

A lot happened following publication of that book. In February 2011 I was engaged by an international consultancy on the new Tripoli International Airport project in Libya, when a major 'civil commotion' broke out which was well documented in the international media. After almost two days of tense waiting, we were finally evacuated by the British government. Since then, my work has taken me across the MENA region to Oman and the major oil and gas exploration company in the country. In late summer 2011, Routledge suggested this related textbook on an *Introduction to Estimating for Construction* and I readily accepted for the reasons above.

The contents and structure of this book are slightly different from traditional texts on construction cost estimating. I make no apology for this, and the main reason is that the industry, both internationally and in the UK, has changed quite considerably over the last twenty years – the builder is invariably involved much earlier in the process, which not only affects procurement issues but also cost estimating, since construction companies have a much greater depth of cost information at their disposal. Additionally, the industry has become 'drier', with a greater proportion of work being manufactured off site, ranging from roof trusses to architectural finishings and façades, and even to entire modules or pods which can be lifted into position and only require to be plugged in on site. This reduces the site-based 'wet' trades with their consequent higher levels of wastage and lower levels of quality control. Therefore, this book is designed to integrate the concept of early contractor involvement, whilst still covering the fundamentals of construction cost estimating.

Construction contracting is still a very commercial activity and also one of the last industries where negotiating the lowest possible cost whilst transferring more and more risk to the contractor is still considered as normal practice and the 'thing to do'. The practice of reverse auctions has even accelerated this to new levels. If, after studying this book, together with *Introduction to Building Procurement*, there is a greater realisation that you get what you pay for, together with an understanding of the tensions within the time-cost-quality triangle (Figure 5.2), then it has all been worth it.

Brian Greenhalgh

Acknowledgements

I would like to express my great thanks to the following colleagues and friends who have helped in the writing of this book by providing real life examples to be used, or reading over the material and offering practical advice where they thought it was clearly required.

David Monaghan, Commercial Director of BAM Construction Ltd, North West England
Steven Bone, Director of the Capital Allowances Partnership Limited, UK
Zane Heasman, Cost Manager, Petroleum Development Oman

1 Introduction to estimating construction work

The CIOB Code of Estimating Practice defines estimating as 'the systematic analytical calculation of projected construction and overhead costs for a contract' and is therefore concerned with the process of anticipating the costs of construction work before the work is actually carried out.

There is also the commonly held confusion between what is meant by an 'estimate' and a 'quotation' and the difference is important in tendering procedures. An estimate provided by a builder, or any other service provider for that matter, is not a guaranteed price for the work; it is merely an estimate which could go up (or down, although this is unlikely) depending on the actual costs incurred during the implementation of the work. As this uncertainty is generally unacceptable to most clients, they will invariably ask for a 'quotation', which is a fixed price for the works and is normally legally binding providing the scope of works does not change. Chapter 7 explains the procedure in converting the contractor's 'estimate' to a 'quotation' to be formally submitted to the client. In this respect, the terms 'bid' and 'tender' are interchangeable with 'quotation'.

In terms of estimating the anticipated costs of construction work, there are three basic techniques in common use and this book is designed as an introduction to the principles of estimating using all three techniques, including how they interact with each other at the various stages of the construction project, from feasibility through the design stages and finally during construction itself.

1.1 Approximate estimating and resource-based estimating

First, the costs of a future building can be estimated by a comparison with previously constructed similar buildings. Clearly, there will be differences between the two, mainly in terms of size, location, quality of materials used and the time difference between the respective dates of construction. This comparison will be made in terms of the total building and is generally given as a cost per square metre of the gross internal floor area (GIFA) of the proposed building.

Second, the costs can be estimated by splitting the building down into its component 'elements' (see Table 5.5 in Chapter 5 for a list of these elements) and comparing the cost of these elements with those of similar previous projects. This is a slightly more accurate technique than the first and therefore requires more information about what is going to be built (i.e. an outline design – see Chapter 6 for a discussion of estimating techniques during the design development stage).

These two techniques are often collectively known as 'approximate estimating' and should be accurate at this stage to within ± 10 per cent of the eventual lump sum tender price (which may not necessarily be the final cost of the building, as we shall see).

Third, when detailed design information is finally available, the resources required to construct the building (i.e. labour, materials, plant and equipment, as well as site-based overheads or preliminary costs) can be firmly established and priced. The aggregate anticipated costs of all these resources will be the estimate for the total project. However, the estimate established by this method will only consider the 'direct construction costs' and takes no account of the contractor's general company overheads and any required profit, which are added later by the construction firm (contractor) when converting the estimate to a tender to be submitted to the client. Furthermore, the first two techniques take their data from previously constructed projects, and these will have included the overheads and profit and are also based on contract prices which have previously been submitted in competition (known in cost estimating as 'tender prices') whereas a pure build-up of resource costs does not include any competitive element (known in cost estimating as 'construction prices'). As if that is not complicated enough, if a bill of quantities is used as a project pricing document, the resource costs would need to be converted from the costs of an activity to unit costs of each item in the bill of quantities, which may seem on the face of it to be an unnecessary burden on the tendering contractors.

The need for accurate estimates before the client makes a commitment to go ahead with a project is clearly obvious. The client needs to know the amount of funds they should set aside for the project and, if possible, what their likely monthly outgoings will be during the construction stage of the project. Many corporate clients will prefer the certainty of cost over the actual cost itself, which means that they prefer to be certain that their project will cost, say, £1 million rather than receive vague assurances that the cost *may* be as low as £900,000, as their corporate budgets are set well in advance of the actual expenditure and having to find extra funds may prove both difficult and embarrassing. For this reason, the traditional lump sum competitive tendering approach to construction procurement is still the most common form of procurement in the industry (lump sum tendering still accounts for over 70 per cent of construction work in the UK by both number and value of contracts and an even greater proportion of international contracts). Even if the main contract is let on a different approach, the work packages or subcontracts are mostly still arranged on a lump sum competitive tendering basis and many subcontractors still prefer this approach to the 'new-fangled' ways of partnering and framework arrangements. Paradoxically, as mentioned above, a lump sum contract still does not guarantee that the final cost of the works (also called 'out-turn costs') will bear any resemblance whatsoever to the lump sum tender figure.

Competition therefore remains the fundamental requirement on most construction projects throughout the supply chain (i.e. supplier→subcontractor, subcontractor→main contractor, main contractor→client) thus making the estimating function the most critical part of an accurate and safe tendering process.

Since the mid-1990s, the construction industry has been under considerable pressure to innovate and adapt, brought on by:

a) clients becoming more 'customer focused' and requiring the same approach from their suppliers and contractors
b) government reports such as the Latham and Egan Reports requiring structural changes to the industry

c) economic recessions which reduce workload, thereby requiring firms to become leaner and more cost efficient
d) the various banking crises, leading to more difficult debt financing and business loans.

All of this has meant a necessary and predominantly welcome change in procurement practice away from purely lowest price to a more holistic 'best value' as the main selection criterion. This has also led to an increased demand for early involvement of the contractor in the design development, as they are recognised as having the necessary skills, knowledge and expertise to contribute towards buildability, value engineering, sequencing, programming, scheduling and whole life cost assessments. Contractors' proposals for all of these issues are often required as part of a detailed tender submission by contractors, which gives clients and their advisers the means to select contractors on more criteria than just lowest price. This has consequently changed the structure of the tendering and bidding process from a linear and end-on design→documents→tender→analysis process to full-blown contractors' submissions incorporating design, planning and logistical proposals together with detailed costs. This places far greater demand on the tendering contractors' estimating process in terms of the quantity and quality of the technical inputs required. These changes generally leave the core of the estimating function relatively unchanged, although they may alter the location and the significance of estimating in the overall business process. Many contractors have taken this opportunity to develop their estimating departments into more holistic 'bid management' departments (see section 1.4) incorporating estimating, design analysis, risk management and tender strategy, which now take the lead in preparing tenders. In fact, the term 'tender' is slowly being replaced by the term 'request for proposal' (RFP), which implies a greater input from the tendering organisation than merely a single figure response to tender documents and is also the common term used in other industries as well as in the construction industry internationally.

1.2 Good practice in estimating

When preparing a tender/bid/proposal (whichever term is being used on the project), the contractor is usually responding to an invitation to tender (ITT) or a request for proposal (RFP) from the client or their consultants and the client will normally have a list of selection criteria and scoring matrix in order to assess each bid as fairly and comprehensively as possible. Therefore, establishing what the client wants from the project is of critical importance at the outset and any tender submission documents should demonstrate very clearly how these client requirements will be met should the tendering company be eventually awarded the project.

Each tendering contractor should undertake a thorough examination, both of the tender documents themselves and also the physical site for the construction works additionally a historic desktop study may be necessary to establish previous occupancy of the site and any likely contamination issues. Whether or not the design has been progressed to an advanced stage, a design review is also essential in order to fully understand the requirements, make preliminary assessments regarding method and sequencing of the works and also to establish if there are any inconsistencies or ambiguities between the tender documents themselves, e.g. between the drawings, specification and bill of quantities. Any questions arising from this review should be forwarded as soon as possible to the client or their consultants via a formal tender query list and, if time permits, to arrange tender clarification meetings with the design consultants to discuss and address these and any other queries. Obviously, the client and consultants will need to be fair to all tenderers and they are required to treat each tenderer in exactly the same way under the Code of Tendering Procedure which is appropriate to the project. This may mean that tender clarification meetings are attended by all tenderers, which may restrict the discussions and questions as no one firm will wish other competitors to gain any advantage or be seen to ask silly questions; however, these meetings are rarely if ever used in the UK, although they do take place many other countries. Having said this, any inconsistencies or ambiguities in the tender documents may provide fertile ground for applications by the contractor for additional costs, variations or claims during the construction process, so if an individual tenderer notices any such inconsistencies in the tender documents, they may well keep the knowledge to themselves in the hope of profiting later on. Shame on them. Also, during periods of recession, when tenders are at their most competitive, a contractor may be tempted to submit a tender at less than actual cost, which hopefully has a better chance of winning the bid, in the expectation of recovering their profit margin through variations and claims during the construction stage of the project. This is referred to as 'suicide bidding' – see section 7.3.2 for further discussion on this issue.

Depending on the tender documents and procurement route chosen, the contract terms and conditions may also include a 'value engineering' clause, whereby the contractor is allowed to suggest alternative products or techniques at tender stage, which reduce the capital cost whilst still maintaining the same functional performance and/or whole life costs. In this case, it is easy for the specific omission to have more far-reaching consequences than the estimator may be able to foresee, therefore a full design review is a necessity for the tendering contractors. Even during the construction stage, when the contract has been signed and the scope of works has been defined, there may be a 'value engineering incentive clause' in the contract, whereby the contractor is encouraged to suggest improvements in cost and/or efficiency of the design, receiving a proportion of the cost savings as an incentive for their efforts. In these situations, the good estimator will tread very carefully in order to ensure all impacts have been considered.

Also, the concept of a 'lump sum' contract may have different meanings in different countries and in different industries. In the UK, a lump sum contract usually means that the tenderers submit a single figure at the tender stage which is hopefully a total of all the items in the pricing document (usually a bill of quantities). Any differences to the quantities will be subject to a variation order from the client (depending on the terms of the contract). However, in many overseas jurisdictions, and now creeping into the UK construction industry, the bills of quantities are at the 'contractor's risk' and it is not unusual for tendering contractors to completely re-measure the project if they suspect the bills of quantities to be inaccurate. This clearly adds extra overhead costs to the tendering contractors, which are only recovered if they are awarded the contract. See section 7.4.2 for a discussion of tender performance and bidding strategy and how contractors can minimise these overhead costs by increasing the efficiency of their tendering process.

The estimating/bidding department therefore needs to draw in skills from across the company, as the tender or proposal which is submitted must be a submission from the company as a whole, not just the estimating department. If the tender is successful, it is not the estimators who will carry out the work – the project will be handed over to the operational departments of construction and contract management and they naturally need to understand how the bid/proposal has been put together, what methods and sequencing have been assumed and what cost budgets they will have to work to, including direct costs, preliminaries and site overheads, temporary works and the required head office overheads and profit. Consequently, in order to complete the contract successfully and profitably, the estimate and tender must be calculated in accordance with company policies which are understood and accepted by all the other departments and clearly demonstrate how the contract would meet the company objectives and business plan relating to work practices and profitability.

1.3 Contractor selection criteria and prequalification

Many construction clients are becoming increasingly more sophisticated and knowledgeable about the construction industry, especially those clients with a significant construction workload, who are often termed 'sophisticated' clients. Also, depending on the choice of procurement route and the point in the procurement process that the contractor is appointed, as stated previously, the criteria for selection may, and often will, be based on many other factors than just cost alone. Issues such as the experience of the proposed team, choice of construction methods, project sequencing, health and safety, duration and any value engineering proposals may be just as important as cost in the client's opinion. Clients and their consultants often use selection panels, interviews and evaluation matrices to assist in contractor selection for a particular project or series of projects. These interviews and presentations by tenderers to the client and consultants are often termed 'beauty parades' in the industry and care must be taken that objectivity and sound technical proposals are not reduced to secondary criteria by smooth talking salespeople, 'business development managers', slick presentations and 'death by PowerPoint'. Most sophisticated clients and experienced consultants are usually well practised in evaluating tender submissions and will be unimpressed by volumes of standard or photocopied information or presentations which do not address the needs of the particular project. Construction is a practical industry and clients, who are investing significant sums of money, will want to see the whites of the eyes of the people who will be working in the senior positions on the project.

1.3.1 Prequalification for a tender list

Best practice normally requires that a tender list for a construction project should contain no more than eight companies, depending on the method of procurement chosen. The maximum size of a tender list for the various procurement routes in the UK is advised in the appropriate Code of Procedure for Tendering and limiting the size of the tender list clearly has financial benefits to both the client and contractors:

a) It increases the chance of success of each tenderer as the competition is limited.
b) The tenderer is therefore more willing to devote time and effort to the tender, thereby increasing the likelihood of a competitive price.
c) It reduces the cost of sending tender documents – although many tenders are now sent out only in electronic format.

d) It reduces the time taken (and therefore cost) for tender evaluation by the client or consultant of all the submissions.

Many clients have the mistaken idea that the larger the tender list, the more competitive the bid price. This is a specious notion for the reasons set out above and it is important for clients to be advised of the optimum size of tender list for the project under consideration. This equally applies to work package contractors when the main or management contractor is appointed early in the project.

Additionally, when a tender is required to be advertised publicly as an open tender, for example, when the project falls within the OJEU guidelines, the client will often go through a prequalification procedure to decide the final tender list and consequently who will receive the detailed tender documents. This prequalification procedure is intended to ensure only those contractors who have the required skills, competence, experience and resources are allowed to submit full technical and commercial tenders for the project, in limited competition.

The 'invitation to prequalify' will normally include:

1 A general description of the scope of works of the project.
2 A general description of the prequalification process and instructions for submission.
3 The procedure for any queries during the prequalification tender process.
4 A questionnaire to assess the financial strength of the tendering company to execute a contract of the type and magnitude envisaged.
5 A questionnaire to assess the experience of the company as a main contractor of similar projects to that envisaged. The tendering companies will be required to submit a list of previous projects undertaken, preferably with client references.
6 A questionnaire to assess the tendering company's key personnel who have the level of experience required by the client. A list of key roles would be supplied by the client and will normally include the project manager, commercial/contract manager, construction manager, programming manager/project control manager together with HSE and QA managers. The tenderer will be required to provide CVs of staff proposed to fill the above posts.
7 A questionnaire to establish the company's corporate HSE policies and procedures together with evidence of past HSE performance.
8 A questionnaire to assess the company's quality management system, which should be in accordance with appropriate international standards. Copies of all quality assurance certificates should be attached to the contractor's submission.

When the contractors have submitted their prequalification material, the client and/or their consultants will evaluate the submissions and complete an evaluation matrix similar to Figure 1.1. This ensures that all contractors are treated fairly and equitably and that the final results for each contractor reflect the importance that the client puts on the various criteria.

Where there are a group of clients with similar projects within a geographical region or industry sector, it may be beneficial to centralise the contractor registration and prequalification procedures using a consultancy company specialising in this type of business. This is often known as 'community prequalification' and is common in

Project No.
Project name:
Date:

PREQUALIFICATION SCORING EVALUATION
WEIGHTED SCORING EVALUATION

Form	Title	Form sub-heading	MAX SCORE	CONTRACTOR A	CONTRACTOR B	CONTRACTOR C	CONTRACTOR D	CONTRACTOR E	CONTRACTOR F
		COMPANY INFORMATION							
		Registered full name of the company							
		Date established							
		Date established in the region							
		Address							
		Telephone and fax number							
		Main contact person							
A	FINANCIAL CAPABILITY								
	1	Is contractor under any financial claim of any size?							
	2	Contractor to provide valid licence to operate							
	3	Provide copies of published reports and audited accounts for last 3 years							
	4	Details of any JV partner for this project							
	5	If the contractor is a subsidiary, details of parent company to be provided							
B	EXPERIENCE								
	1	Number of similar projects completed in region over last 5 years							
	2	Number of similar projects completed in locality over last 5 years							

Figure 1.1 Example of a weighted scoring matrix for contractor prequalification *(Continued)*

Figure 1.1 (Continued)

<table>
<tr>
<td>3 Provide names and contact details of at least TWO client references.
4 Provide details of site which may be visited as part of prequalification procedure

C RESOURCES
1 Provide CVs for key personnel (may be on ‘or equal’ basis)
2 Provide typical example of project organisation chart for similar recent project. Chart to include names of senior project personnel
3 Confirm ability to mobilise within 30 days of contract award
4 Confirm details of trades to be covered internally and trades to be subcontracted
5 Provide details of subcontractor verification procedures

D HSE
1 Provide details of corporate HSE management system / operations manual
2 Provide details of environmental assurance system equal to ISO 14000
3 Have all key personnel received formal HSE accreditation?
4 Provide details of subcontractor verification in HSE issues, especially HSE competence and records</td>
<td></td>
<td></td>
<td></td>
<td></td>
<td></td>
<td></td>
<td></td>
</tr>
</table>

5 Provide details of system and upkeep of PPE - standard and specialised 6 Provide details of HSE records of performance and incidents for last five years 7 Provide example of traffic management plan for a project site 8 Provide details of written policy on HSE auditing 9 Any legal proceedings taken against the company or parent company regarding HSE issues in last five years? E QUALITY 1 Provide details of QA/QM system to latest ISO 9001 2 Provide copies of typical project Quality Plan and Quality Control Plan 3 Provide details of subcontractor approval process 4 Provide details of any specific measures contractor would take to ensure overall client requirements are fully met							
TOTAL							

process engineering projects such as oil and gas installations where the prequalification criteria of the various operators (clients) have a high degree of similarity.

The advantages of community prequalification include:

- savings in time and money from repetitive prequalification questions to suppliers as the information is available from consultants
- more easily allows new entrants to become prequalified
- a consistent, structured and transparent format of contractor registration and selection
- access to shared information with the industry and improving performance with suppliers
- the ability to interrogate contractor information to produce short lists
- up-to-date information on suppliers as consultants continuously update the lists.

It is important for the estimating and bid management departments to note that prequalification submissions should never include pricing or cost information related to the project. This is not a tender (yet).

1.3.2 In-country value (ICV)

Many developing countries are now looking very seriously at their supply chain and the amount of expenditure throughout the supply chain which is retained in the country. This has become even more important in countries which have experienced social unrest and protests during 2011 and 2012, where a high proportion of the local population are unemployed but there is major capital expenditure on infrastructure projects which predominantly goes to international companies. These companies in turn will usually purchase materials and labour from outside the country as well as repatriating their profits to their own countries of origin.

A definition of ICV is 'the total expenditure (by a client) retained in-country that can benefit business development, contribute to human capability, development and stimulate productivity in the local economy'. This includes the procurement of materials and services from local sources, the hiring of local personnel and training to improve their capacity and capability to perform the various project functions.

This initiative, whilst clearly being a worthwhile project from a host country's perspective, also means that the tendering contractors will not be able to resource the projects as they would wish with the most cost-effective resources at their disposal and are also required to submit themselves to a rigorous prequalification or validation procedure, which adds more costs to their company overheads.

It is likely that, in the future, more developing countries will engage with this initiative in order to speed up the overall development of their own economies. There is also an interesting argument that the developed economies in Europe and North America may take up some of the in-country value policies in order to help in the recovery from the sovereign financial crises in 2011 and 2012.

1.3.3 Selection of the contractor

Returning to more project-specific issues, for the more complex proposals, and certainly where there is an early appointment of the contractor, i.e. before the design has been completed and the contractor is required to contribute to the design, the formal tender submission is often split into two parts:

a) technical proposals
b) commercial and financial proposals.

The two proposals would be submitted separately and the technical evaluation of the contractor's proposal is normally carried out first, preferably before the commercial bid has been opened. If a particular technical proposal is rejected for any reason, that company's commercial proposal will/should not be opened, thus ensuring that only the commercial proposals of technically approved and compliant bids are considered. It is relatively easy to imagine that a company who puts in a poor technical proposal may well be able to submit a much-reduced commercial and financial bid and thereby have a greater chance of winning the project if cost is the only or major criterion. It is also an option to open the bids the other way round, i.e. to open the commercial bids first and then technically evaluate the lowest bid – if that is not acceptable, the technical bid of the next lowest would then be considered, and so on. Again, it is easy to see that a cost-conscious client may well exert some pressure to accept the lowest bid, even though the technical content is poor. For this reason, the evaluation of technical bids before commercial bids is the safest option. The sour taste of poor quality lingers long after the sweet smell of cheap price has gone.

Of course, this only applies to project proposals where the tendering companies are requested to submit technical solutions based on their own design skills or technical know-how, such as EPC contracts where the contractors are required to base their tender on a 'front-end' schematic design only. See sections 3.3.1 and 3.3.2 for further discussion on this issue. This would also apply if there is a significant value engineering requirement at this pre-construction stage. For projects where the design is complete and all tenderers are required to submit compliant tenders based on the client's tender documents alone, there is little need or benefit for separate technical and commercial bid openings.

An example of this from the author's own experience is in the electronic security and safety subcontract package for an international airport in the MENA region. Because of the size and complexity of international airports, the design was completed several years before the electronics packages were due to be installed and technological advances in this area are rapid. Additionally, international standards for airport security are continually changing and becoming more rigorous. Therefore, it is important to utilise the skills of the specialist subcontractor and for them to propose solutions which comply with the current international standards and regulations as well as using the latest technology. Each tenderer may propose something slightly different and the client team must be able to assess whether the various proposed solutions meet their necessary functional requirements as well as ensuring that they do not cause coordination or interface problems with other packages.

1.3.4 Technical and commercial evaluation report

When the evaluations have been completed, an evaluation report is normally submitted to the client's decision-making body – usually termed a tender board or tender committee. The contents of this report would include the following:

A Introduction
  A1 Opening of tender
  A2 Methodology
B Compliance with the requirements of the tender documents
  B1 Conformity and completeness of tenders
  B2 Proposed submittals, programme management, technical solutions etc.
    Contractor A
    Contractor B
    Contractor C
    Etc.
C Technical evaluation
  C1 Evaluation of technical submittals by individual tenderers
  C2 Post-tender technical clarification meetings (if any)
  C3 Further design clarifications following meetings in C2
  C4 Technical rankings of tenderers prior to commercial evaluation
D Commercial evaluation
  D1 Completeness of offers
  D2 Currency of bid
  D3 Delivery basis of quotation
  D4 Minimum delivery time
  D5 Validity of offer (compliant with tender documents?)
  D6 Submission of bid bond
  D7 Total value of bid
E Conclusions and recommendations

Tender committees are often very powerful within the client organisation, as they are responsible for the approval and award of major capital expenditure. For this reason, the members of the committee are experienced and senior executives of the company and the committee may also include representatives of shareholders, stakeholders or joint venture partners. The terms of reference of the tender committees are normally to approve or endorse recommendations for contract award and at other major milestones within the procurement process. Therefore, the recommendations to these committees should be backed up by full and complete evaluations and detailed recommendations with a clear and transparent audit trail.

Although the tender committee would approve or endorse the placing of capital projects, the letter of intent, letter of award and formal contract would be issued by the functional department responsible for contract management and administration. For further discussion in this area, the reader is advised to consult a good textbook on building procurement (see the bibliography at the end of this book).

1.4 Bid management

When an invitation to tender is received by a contractor, it is vital to establish a formal procedure to process the tender (bid) effectively and efficiently. This procedure is generally known as the bid management process and would cover the following areas:

- *RFP (tender document) analysis and initial evaluation/planning*. The bid management department would take the leadership of the initial stage including a detailed

RFP analysis and assessment of client requirements; documenting the client's critical success factors; a response plan and agreeing assignments; development of flow-down subcontract packages; planning and agreement of additional material for submission; leadership of any pre-submission client presentation/clarification meeting.

- *Project management*. Coordination of the bid response: management of subcontractor submissions, production of proposal documents by the required deadlines.
- *Commercial leadership*. Analysis of subcontractor bids to ensure coordination with the main contract, collating and compiling into the main bid proposal. Additional material for bid submissions, such as company financial data, information on key personnel, internal presentations to senior managers and signing-off of submissions.
- *Internal leadership*. Scheduling and managing the internal bid progress and approval meetings and liaison with other offices, departments etc. as required.
- *Tender deliverables*. A schedule of what has to be provided and returned with the tender.

Clearly, the effect of all these new procedures on the estimating process will be to greatly increase the number of inputs and outputs, making the expanded estimating function a much more strategically important area of the organisation.

Another aspect of seeking best value rather than lowest price is an increase in the attention paid to the supply chain involved in tendering and delivering contracts. Specialist subcontractors have a critical role in the modern hi-tech industry where client requirements for specialist MEP, HVAC and telecommunications services together with more rigorous environmental and low carbon regulations means that the actual building is often seen as a mere envelope for the complex technology inside. Therefore, specialist contractors are seen as increasingly relevant in the demonstration of best value, as it is their resources that will be applied and which will be largely responsible for realising the benefits of innovation and technological development. The need for assurance of high-quality delivery from subcontractors has encouraged many main contractors to enter into formal and informal partnering arrangements with specialist contractors who can demonstrate a good track record. This may have an effect on the procurement approach for client organisations, as they are aware of the need to engage high-quality specialists with their own quality assurance procedures and guarantees of best value. The more sophisticated clients may also contract with specialist firms themselves, requiring the main contractor to liaise and coordinate their work.

The estimator's relationship with specialist contractors during the tender period must therefore become much more intimate, with more collaboration on design development, construction methods, sequencing and cost estimating – traditionally, the complex services were seen as a 'black box' with few people outside the specialist firm understanding the technology. The need to work in this way will also impose some restrictions on the way in which some estimators have, historically, managed the process of procuring competitive subcontract quotations.

Tendering is an expensive business for contractors and the costs of tendering rise in proportion to the requirements of the tender documents; therefore it is important for a contractor to maintain a good success rate in the tenders they submit. Tender performance and bidding strategy will be discussed more fully in Chapter 7 but the estimator should constantly look for cost advantage throughout the scope of the bid, especially regarding subcontractors, design simplifications and the choice of methods and reduced non-productive down time for plant and equipment.

1.5 Risk and value management

The management of risk is a key thread which must be part of the estimating process at each stage of the bid management, along with the need to make sufficient allowance for health, safety and environmental (HSE) legislation. A well-produced risk and opportunity register will carry forward into the construction phase and will become a key tool for the project manager.

There is always an element of risk in anticipating the future and nobody can be entirely sure that the decisions and judgements that are made at the estimating stage will be correct at the time when the project is actually being constructed, which may be more than a year after the estimate has been calculated. Mistakes and errors are also possible either through the use of incorrect data or just basic arithmetical errors in adding up the totals. Therefore, estimates cannot be guaranteed to be error free but good practice, professionalism and robust risk management strategies should help to keep errors and their financial effects to a minimum. In a competitive tendering environment, it is often said that contractors win tenders because they have missed something out or made a mistake which reduces the total tender figure to below that of their competitors. This is clearly a cynical view, but it does happen in practice, especially when the period given for estimating and tendering is very tight and the tendering contractors may not have had the opportunity for full and rigorous bid management.

1.6 Health, safety and environment (HSE)

Health, safety and environmental issues have become increasingly important in the construction industry over the past 15 to 20 years, driven by increasing awareness of environmental considerations and legal requirements regarding health and safety in the workplace, and especially on construction sites, which were (and still are) notorious for having poor safety records.

During the estimating and tendering stage, it is important for the potential contractor to gather as much health and safety information about the project and the proposed site as possible. Information available at this stage should be used to make allowance for the time and resources required to deal with particular problems during construction. Sources of information would include the client, the design team, the tender documents, any specialist contractors, consultants and suppliers already appointed, together with the HSE (confusingly this is the same acronym, which this time stands for the Health and Safety Executive, part of the UK government) guidance and British, European or International Standards documents.

The tenderer should find out about the history of the site and its surroundings, such as any unusual features which might affect the work, or how the work will affect others in the vicinity. Particular attention must be paid to contaminated land or dangerous materials such as asbestos, together with overhead power lines, underground services, unusual ground conditions, any public rights of way across the site, nearby schools, footpaths, roads or railways etc.

In the UK, where the Construction (Design and Management) (CDM) Regulations 2007 apply, much of this information should be found in the pre-construction-stage health and safety plan, which is required to be produced by the client (note that the *client* is legally required to produce this information, but in practice it will be written by their consultants). All tenderers must ensure its contents have been taken into account before tenders are submitted. Where CDM does not apply, gathering such information is still important as there is still a legal requirement in most countries to create a safe workplace for employees and others working on the site.

When estimating costs and preparing the programme schedule, tendering contractors are required to consider any particular health and safety hazards associated with the work and suitable allowances should be made in the tender price. The project will run much more smoothly, efficiently and profitably if hazards have been anticipated, planned for and controlled from the outset. Having to stop or reschedule work to deal with emergencies clearly wastes both time and money.

When materials are bought, or plant and equipment is hired, the supplier also has a duty to provide certain health and safety information. Estimators should ensure that this is obtained and taken into account. If it is a particularly important potential hazard, it may be necessary to consider using a specialist who is familiar with the necessary precautions, but in all cases, the tenderer should carry out an assessment of the health risks arising from substances or equipment; and act on the findings, e.g. by eliminating harmful substances where possible, or by using a less hazardous method of work or providing training on the safe use of the material or equipment.

When tender programmes are being prepared, contractors must consider whether there are any operations that will affect the health or safety of others working at the site. For example, in terms of site access – which trades and groups of workers will need to go where and when? The construction programme should also ensure that everyone who needs to use an item of general plant has sufficient time allocated to do so. It hopefully goes without saying that the access to the item of plant must be safe and suitable for their use.

Main contractors should always discuss any proposed working methods and method statements with subcontractors before letting those subcontracts. The main contractor should find out how they are going to work, what equipment and facilities they are expecting to be provided and the equipment that they will bring to the site themselves. They must also identify any health or safety risks that their operations may create for others working at the site and agree appropriate control measures. For example, many contractors only allow 120 V power tools on site for reasons of electrical safety and also do not allow the use of mobile phones, portable radios or personal stereos/iPods etc. during work time on site.

Obtaining health and safety risk assessments and method statements will help to decide what plant and equipment will be required and check that it will be suitable.

If there are significant hazards, or the site is particularly complex, the site will require formal procedures for emergency and rescue. The tendering contractor must decide what equipment will be required, its location on the site and who is trained to operate it.

1.7 Supply chain management

Businesses in all industries realise that working in a positive and collaborative way with other companies that supply goods and services, or with customers to whom you supply goods and services, is part of normal business practice and good relationships make for good long-term profitability. Construction companies that work in this way have been seeing the benefits both for themselves and their clients. Supply chain management (SCM) is merely a formalisation of the contractor's business processes to ensure that these arrangements work in practice.

The 'supply chain' is the term used to describe the linkage of companies that convert basic raw materials, labour and equipment into a finished product for the client or eventual end-user. All project participants, such as the client, main contractor, designer, other consultants, subcontractors and suppliers are therefore part of a supply chain and because of the project-based nature of construction and the way that procurement normally operates, they may be members of different supply chains on different projects for the same client.

According to Constructing Excellence, products and services provided by the various companies in a construction supply chain can typically account for about 80 per cent of the cost of the project. The way in which these products and services are procured and managed may have a considerable effect on the outcome of the project, in terms of profitability for the contractor and also for the way in which the completed facility meets the client's expectations of time, cost, quality and functionality.

Each company in the supply chain therefore has their own internal customer, i.e. the organisation which directly uses their products, or put another way, the next company in line. However, a fully integrated and properly managed supply chain will have the objective of all members both working towards and understanding the interests of the project client and the end-user (who may be different).

In traditional contractual relationships, the companies may only be linked by contracts that have been based on lowest price against fixed specifications. The supplier is required to deliver the specified product or service as cheaply as possible. There is therefore little or no incentive to work in the client's interest, or to use any specialist knowledge to achieve better value for the client. Fortunately, modern procurement methods are moving towards the appointment of integrated supply chains where all the parties have a long-term objective to work together to deliver better value to the client and the end-user.

Consequently, the benefits for individual companies in long-term SCM include:

- reduced real costs, whilst still retaining profit margins
- more incentive to remove waste from the process
- greater certainty of out-turn costs and quality of installation
- better value to the client.
- more repeat business with key clients and supply chain partners
- greater confidence in longer-term planning.

The benefits for end-users and project clients will include a more responsive industry able to deliver facilities that meet user needs in a better way, delivered to time and cost with minimal defects. This in turn creates higher customer satisfaction levels and, hopefully, an improved reputation for the industry.

All construction companies have lists of preferred subcontractors and preferred suppliers. Therefore, there are established relationships already that the contractors can build upon. To fully realise the potential of SCM, the contractor and their supply chain would be able to develop an 'offering' for clients based on the better value of appointing a ready-made full supply chain, which is of course one of the major selling points of early contractor involvement in construction projects. Having established relationships will mean that all parties will better understand the processes with reduced opportunities for ambiguities and disputes.

Whether the supply chain is being established for a single project or by a group of companies as a marketing tool, the following principles apply:

KISS – keep it simple sweetheart

Don't try to be too clever with the agreements between the parties. Start by establishing relationships with those suppliers and subcontractors who are critical to the delivery to the client or project. These are the strategic supply chain partners or 'first tier suppliers'. It is vital that time and care is taken to establish which companies fulfil the criteria of strategic partners and that they have similar serious interests in developing long-term relations. A successful supply chain of first tier suppliers is a manageable objective and in time each of these suppliers will have similar chains.

Evaluate potential first tier suppliers in the following areas

This is effectively a 'prequalification' of the suppliers; so much of the discussion in section 1.3.1 is relevant here. In particular, the following areas should be evaluated:

- the history and success of the existing relationship
- technical capability and industry reputation
- design capability and innovation record (especially for involvement in the project design stage)
- size and market position
- management style and ability to integrate with other companies.

What is being sought is a partner capable of reliably supplying quality products and services at competitive prices on a long-term basis. A successful partnership will deliver mutual commercial benefit through greater success in the market, based on delivering better value to increasingly satisfied clients. All parties in the supply chain must be committed to working for the long term on the basis of continuous improvement and innovation. If anyone is inclined to quit when the going gets tough, the supply chain is likely to fail.

Involve the designers

Early involvement of the contractor requires a design input to the project; therefore the supply chain partners must include design professionals (e.g. architect, structural engineer, services engineer). However, where the supply chain is established to deliver a specialist part of the facility, the design function may be embedded within one of the supply chain partners. In either case, the designer's role is central to delivering optional functionality, value for money, whole life costing and safe construction using optimal labour resources and minimum waste.

Manage the costs

Cost management is of course central to successful collaborative relationships as all parties still need to make a profit. Even the most hardened clients must accept that they will obtain best value if the supply chain's margins are offered some protection, as they are all commercial enterprises. By this recognition, the supply chain can focus on delivering value to the client rather than using its efforts to protect margins or chasing claims.

Even where the supply chain cannot get such agreements from a client, it is still essential that costs are understood and managed. The principle of sharing risk and reward underpins the whole process of collaborating for mutual benefit. Passing all risk down the supply chain is an easy option and will not lead to the lowest cost and certainly will not lead to best value for the client in the final analysis. A painshare/gainshare arrangement will often be helpful in this regard.

In summary, therefore, the key objective of supply chain management is to offer better underlying value to a client than the competition, which is done through a combination of

- defining client value
- establishing supplier relationships
- integrating activities
- managing costs collaboratively
- developing continuous improvement
- mobilising and developing people.

1.8 Project planning and method statements

If 6 to 8 contractors on a tender list receive identical tender documents which set out the detailed design of a construction project in a particular location at a particular time, then why should the tender prices be so different from each other? Considering the different components of the tender price:

- *Material costs.* If the design is completed and the specification of materials decided, the material costs should be broadly similar for all tenderers, assuming that each tenderer has efficient business processes for materials purchase. Therefore the material costs should not create a large differential in the tender costs.
- *Labour costs.* The all-in rate for labour (see Chapter 2 for a detailed explanation of this calculation) should again be broadly similar for companies in a competitive environment, especially where there is free movement of labour between companies, so that a broad equilibrium price is established. Labour productivity (see section 2.2.2) should again be broadly similar for efficient competitive companies. Therefore the actual labour costs should also not create a large differential in the tender costs. However, the labour costs *will* vary according to the method of construction to be adopted by the contractor and there is invariably a trade-off between labour costs and equipment costs as explained below.
- *Plant and equipment costs.* The costs of plant and equipment will certainly vary according to the method of construction adopted by the contractor. There are a considerable number of ways that a construction process can be carried out; for

example, at a very basic level, excavating a trench can be carried out by hand digging, in which case the labour element will be high and the plant and equipment costs negligible (and the time required for the operation will be high). However, this is unlikely to occur on most sites, with the contractor choosing to excavate the trench with a machine that they either own or hire for the project. This will reduce the labour costs, increase the plant/equipment costs and shorten the time for the operation. Therefore, the choice of plant and equipment will have a major effect on overall tender costs and is also closely connected with the choice of method of carrying out the works, established in the method statement.

1.8.1 Contractor's pre-tender programme/baseline schedule

Very early in the tender process, the estimator, or bid manager, will need to establish a tender programme. In many international projects, the term 'programme' is interchangeable with the term 'schedule' and often the first programme/schedule developed for a project is termed the 'baseline schedule' or BLS. This is a North American term and merely refers to the project programme and its revisions. In this North American terminology, the term 'program' does not necessarily refer to the chart which relates one operation with another, but to a series of projects from one client organisation (i.e. a program of projects). Therefore, above an individual project manager would be a program manager. Very confusing until you are used to it!

During the tender process, the contractor will need to draw up a tender programme. This will not usually form part of the contract documents, but will be used by the contractor to establish the expected construction period and the durations for the critical elements, together with the overall sequence for the works and the procurement of materials and subcontractors. If the programme is sufficiently detailed with reference to the contract scope, it becomes of vital importance when the project begins to change as a result of any changes requested by the client, or actions on site which are not within the contractor's control, such as unforeseen weather or ground conditions. This tender programme will also be used by the contractor to show how they had originally intended to sequence the work and consequently, when compared with an as-built programme or progress programmes, shows how much the actual programme has changed from the planned programme (on which the tender was based). This is a very important tool in establishing any claims for loss and expense due to delay and disruption to the works or for attempting to prove an extension of time or acceleration claim. The accuracy of the tender programme very much depends on the effort which was put into it at tender stage and the skill of the planner and estimator. The tender programme and the tender stage method statement should be developed concurrently and taken together; they will give a detailed picture of how the project is anticipated to be built and how long it should take. Figure 1.2 shows an example of a pre-tender programme with critical supervision resources as required during the construction duration.

See section 6.3 for a fuller discussion of the use of programmes/schedules in estimating.

1.8.2 Contractor's tender method statement

A method statement describes the way in which the building or facility is to be constructed. It will mainly be concerned with potential hazards and safety, quality and

XYZ CONSTRUCTION LTD **PROPOSES SINGLE STOREY OFFICES** **TENDER PROGRAMME**

2013

WBS	Activity	Approx Quantity	Gang size	Duration	Jan				Feb				Mar					Apr					May				June			
					1	2	3	4	5	6	7	8	9	10	11	12	13	14	15	16	17	18	19	20	21	22	23	24	25	26
1	Mobilisation	n/a	n/a	2 wks	■	■																								
2	Excavation & earthworks	Values inserted from Bill of Quantities	Dependent on Company policy and work practices	3 wks			■	■	■																					
3	Formwork pad foundations			2 wks						■	■																			
4	Reinforcement pad foundations			1 wk								■																		
5	Concrete pad foundations			1 wk									■																	
6	Drainage trenches			1 wk									■																	
7	Reinforcement ground floor			1 wk										■																
8	Concrete ground floor			1 wk											■															
9	Reinforcement columns GF			2 wks												■	■													
10	Formwork columns GF			2 wks													■	■												
11	Formwork soffit & sides first floor			2 wks														■	■											
12	Reinforcement FF slab			2 wks															■	■										
13	Concrete FF slab			1 wk																	■									
14	External walls			2 wks																		■	■							
15	Flat roof covering			2 wks																			■	■						
16	Windows & doors			1 wk																					■					
17	Internal partitions			2 wks																						■	■			
18	MEP first fix			1 wk																							■			
19	Internal finishings			2 wks																								■	■	
20	MEP second fix			1 wk																									■	
21	Decoration			1 wk																										■
22	External work & drainage			1 wk																										■
	Preliminaries & Site Establishment	Site agent			✓	✓	✓	✓	✓	✓	✓	✓	✓	✓	✓	✓	✓	✓	✓	✓	✓	✓	✓	✓	✓	✓	✓	✓	✓	✓
		Project engineer			✓	✓	✓	✓	✓	✓	✓	✓	✓	✓	✓	✓	✓	✓	✓			✓			✓			✓		✓
		Foreman					✓	✓	✓	✓	✓	✓	✓	✓	✓	✓	✓	✓	✓	✓	✓	✓	✓	✓	✓	✓	✓	✓	✓	✓
		Crane driver												✓		✓	✓	✓	✓	✓										
		Forklift driver										✓	✓	✓		✓	✓	✓	✓	✓	✓									

Figure 1.2 Example of a pre-tender programme

the logistics of the programming and sequencing decisions, what plant and equipment will be used and how the processes interact with each other, including all work carried out by subcontractors. Under UK health and safety legislation, a pre-tender method statement is a legal requirement.

The main contractor's method statement should include:

- definition of the scope and extent of work in each package with particular reference to how safety will be ensured in the potentially hazardous methods of working such as lifting and working at heights or in confined spaces
- assigning responsibility for any work at package interfaces
- ensuring that the separate method statements of package contractors or subcontractors correlate with each other and that taken together they cover all aspects of the work
- submitting the relevant method statements to the client or consultants
- identification of key site-specific issues
- site traffic management, delivery and logistics
- site establishment and services.

A pre-tender method statement will be submitted with a tender bid and a detailed method statement is required to be produced before construction begins and should be prepared with the same care and thoroughness required for any other contract documents, such as specifications, bills of quantities or design drawings.

All contractors are under a legal requirement to conduct a risk assessment of work procedures to be carried out prior to commencement of work, and for the outcome to be communicated to the workforce. Safe systems of work and lifting plans within a detailed method statement are a good way of doing this. The detailed method statement will also describe the project's quality procedures to show that the building is being constructed to the required standards as set out in the contract documents.

In terms of the estimating function, different procedures and work methods will clearly have different cost implications. The contractor will naturally wish to use the most cost-effective method of construction, weighing the overall cost of labour and plant & equipment with the productivity, speed and safety of each method. As mentioned earlier in the chapter, the contractor's competitiveness will be more related to the appropriate choice of construction method than any other issue, including the level of overheads and profit. Therefore, the tender method statement is a crucial document on which the project estimate is based. See Figure 1.3 for an example.

See section 6.2 for a fuller discussion of the use of method statements in estimating.

1.9 Summary and tutorial questions

1.9.1 Summary

The remaining chapters of this book will address the techniques developed by the industry in preparing the various estimates at all of the pre-construction stages. At each of these stages, the client has a decision to make – are they happy with the estimate or not? If yes, the project can progress to the next stage, either to develop the design further, appoint a contractor or start work. At all of these decision/approval points, the cost estimate will be a reflection of separate decisions made regarding

Pre-Tender Method Statement

Contract________________________ Contractor________________

Person completing this Statement______________ Tel./email ______________

Date________________________

Key Activities From Pre-Tender Programme	Plant/ Equipment required	Possible hazards	Recommended Safety controls including PPE	Licences, qualifications or work permits required

Figure 1.3 Example of a pre-tender method statement

quality and time. A higher-quality building will increase the cost, a restricted project duration will also increase the cost, so each project must be assessed on its own merits with an optimum design which is fit for the purpose it is expected to perform and an optimum time which makes economic use of all the resources used, thus creating an optimum cost estimate which should still be within the budget originally set by the client.

It would be useful at this point to summarise some of the different types of estimates used in the pre-contract design stage, which will be further developed in the following chapters. The term 'pre-contract' is used here to refer to the design stage, although with early contractor involvement (ECI), the contract point will be early in the design stage, so the pre-contract and post-contract stages will be different. It all adds to the rich confusion of modern construction procurement!

Design estimates

For the client and their design professionals, the types of cost estimates encountered will run in parallel with the planning and design stages, roughly as follows:

- 'ball-park' or order of magnitude estimates
- preliminary feasibility or conceptual estimates
- detailed estimates developed as design decisions are made
- pre-tender estimates based on detailed plans and specifications.

In the planning and design stages of a project, various design estimates reflect the progress of the design. At the very early stage, the *order of magnitude* estimate is usually made before any design information is available, and must therefore rely on the cost data of similar projects built in the past. A *preliminary estimate* or *conceptual estimate* is based on the concept design of the project at the stage when the basic layouts for the design are known. The *detailed estimate* is more definitive and is made when the scope of work is clearly defined and the design is sufficiently progressed so that the essential features of the project are clearly identifiable. The final *pre-tender estimate* is based on the completed plans and specifications, when they are ready for the client to request bids from interested construction contractors. In preparing these estimates, allowance should also be made for the contractors' overheads and profit.

The costs associated with a project may be disaggregated into a hierarchy of levels which is appropriate for the purpose of cost estimation (roughly equating to a work breakdown structure – WBS). The level of detail of the WBS depends on the type of cost estimate to be prepared. For conceptual estimates, for example, the level of detail in defining tasks is quite broad, whereas for detailed estimates, the level of detail should be much more refined.

For example, consider the cost estimates for a proposed bridge across a river. A feasibility estimate is made for each of the potential alternatives, such as a suspension bridge, cable stayed bridge, cantilever bridge etc. As the bridge type is selected depending on span, ground conditions etc. a preliminary estimate is made on the basis of the preliminary or conceptual design. When the detailed design has progressed to a point when the essential details are known, a detailed estimate is made on the basis of the well-defined scope of the project. When the detailed plans and specifications are completed, a final pre-tender can be made on the basis of items and quantities of work.

As mentioned, for each of these different estimates, the amount of design information available typically increases, thereby increasing the accuracy of the estimate. See Chapter 4 for a more detailed discussion of these different types of estimates.

Tender or bid estimates

For the bidding contractors who are pricing on full design information, a bid estimate submitted to the client either for competitive bidding or negotiation consists of direct construction cost and will also include site supervision, plus a mark-up to cover general overheads and profit. The direct cost of construction for bid estimates is usually derived from a combination of the following:

- direct costs of labour, plant and materials
- subcontractor quotations.

The contractor's bid estimate often reflects the desire of the contractor to win the job as well as the estimating tools at its disposal. The larger contractors have well-established cost-estimating procedures while others do not and rely on publicly available pricing information or a rough estimate of their time and resources. Since only one tendering contractor will be successful, all effort devoted to cost estimating will be a loss to the contractors who are not successful. Consequently, the contractor may put in the least amount of possible effort for making a cost estimate if it believes that its chance of success is not high.

If a tendering contractor intends to subcontract parts of the project, it may request price quotations for the various tasks to be subcontracted to appropriate specialised firms. Therefore, the general subcontractor shifts the burden of cost estimating to subcontractors. If all or part of the construction is to be undertaken by the main contractor themselves, a bid estimate may be prepared on the basis of the quantity takeoffs from the drawings provided by the client or on the basis of the construction procedures devised by the contractor for implementing the project. For example, the cost of a pad foundation of a certain type and size may be found in commercial publications on cost data which can be used to facilitate cost estimates from quantity takeoffs. However, the contractor may wish to assess the actual cost of construction by considering the actual construction procedures to be used and the associated costs if the project is deemed to be different from typical designs. Hence, items such as labour, material and equipment needed to perform various tasks should always be taken into consideration for the cost estimates.

Control estimates

For the purpose of monitoring the project during construction, the contractor will invariably produce a control estimate for their own internal purposes, which is derived from available information to establish:

- budget estimate for financing
- budgeted cost after contracting but prior to construction
- estimated cost to completion during the progress of construction.

Both the client and the contractor must adopt some baseline for cost control during the construction. For the client, a *budget estimate* should be adopted early enough for planning long-term financing of the project. Consequently, the detailed estimate is often used as the budget estimate since it is sufficiently definitive to reflect the project scope and is available long before the more accurate pre-tender estimate. As the work progresses, the budgeted cost must be revised periodically to reflect the estimated cost to completion by a process known as earned value analysis (EVA). A revised estimated cost is necessary either because of change orders initiated by the owner or due to unexpected cost overruns or savings.

For the contractor, the bid estimate is usually regarded as the budget estimate, which will be used for control purposes as well as for planning construction financing. The budgeted cost should also be updated periodically to reflect the estimated cost to completion as well as to ensure adequate cash flows for the completion of the project.

1.9.2 Tutorial questions

1. Why is it important to understand the difference between an 'estimate' and a 'tender'?
2. Discuss the essential difference between approximate estimating and resource-based estimating.
3. Outline the major ways that an estimate can be calculated.
4. For what reasons would a client wish to 'prequalify' contractors on to a selected tender list?
5. Discuss the stages and requirements of the 'bid management process'.
6. Define 'supply chain management' and explain why is it important.
7. Where a client requires ECI, what are the advantages of separating the contractor's technical proposals from commercial proposals?
8. What is the purpose of a tender clarification meeting?
9. Why should the tender programme and tender method statement be constructed before the estimate?
10. What are the major criteria (besides costs) that a client should consider in appointing a contractor?

2 The nature of costs in construction

Cost can be defined as 'that expenditure which has been incurred in the normal course of business in bringing the product or service to its present location and condition' and can be split into two distinct categories. Unfortunately, this can be done in two different ways.

2.1 Types of cost

Costs of construction work can be split into *direct costs* and *indirect costs* and each of these categories can be further split into fixed costs and variable costs. Direct costs consist of those costs incurred on site and directly related to the work being carried out and indirect costs consist mainly of head office overheads and non-site-located service provisions, such as corporate marketing, HR and finance. *Fixed costs* are the same irrespective of turnover and *variable costs* will increase in direct proportion to turnover.

2.1.1 Direct costs and indirect costs

As stated above, costs may be split into direct costs and indirect costs. A direct cost is a cost that is incorporated into the finished work. In construction, all project costs on site can be considered as direct costs, since they are the costs of actually delivering the project. This would include the labour, material and plant costs of the work as well as the preliminaries and other semi-variable costs.

An indirect cost is a cost that cannot be easily and conveniently traced to a particular project. This would include the head office overheads and other common costs or non-project-specific costs.

A particular cost may be direct or indirect, depending on the project and how the company sets up its accounting procedures. While, in the above example, the head office salaries are normally considered as an indirect cost for a particular project, if the company employs a contracts manager to oversee several projects, their salary would be seen as a direct cost to the contracts department, which may be where all project costs are accounted, so it all depends on how the company decides to account for its business operations. Fortunately, this is not in the remit of this book, and we only need to recognise that costs can either be fixed or variable, or they can be direct or indirect.

2.1.2 Fixed costs and variable costs

Costs can be further split into *fixed costs* and *variable costs*. Fixed costs are those which do not change irrespective of the amount of work that is carried out (Figure 2.1). For example, the costs of the head office would be the same if the company had 10 projects or 20 projects. Obviously, in the longer term, these costs can be increased or reduced as required, but in the short term, they are considered as fixed costs or overheads. Similarly, on site many of the general costs are fixed and do not change in accordance with the amount of work being carried out – for example certain preliminary items including the project manager's salary, so in a month where there is little work being carried out, the project manager's salary will still be the same. Notice that we have established two different types of fixed or overhead costs related to construction projects – head office overheads are a fixed cost to the company and must be spread over all the projects that the company is engaged in; general site costs (mostly known as preliminary costs or 'preliminaries') comprise both fixed overhead costs for the project only and variable overhead costs which will be spread over the duration of the project. If the company did not win the tender for this project, they would not incur all of these costs, so at the project level they are fixed costs, but at the company level they may be considered as semi-variable costs. Very confusing – and each company may account for these costs in slightly different ways.

Variable costs, therefore, will change in direct proportion to output (Figure 2.2). This would include the costs of actually doing the work, i.e. the labour costs, material costs and any direct plant or equipment which is used. So a large project will have much greater variable costs than a small project, but the fixed costs will be the same, or only vary marginally or in steps related to the size of the project.

Let us now consider the various costs associated with construction projects. We will first look at what primarily affects costs in construction, and that is the concept of productivity. The various direct costs of labour, materials, and plant and equipment

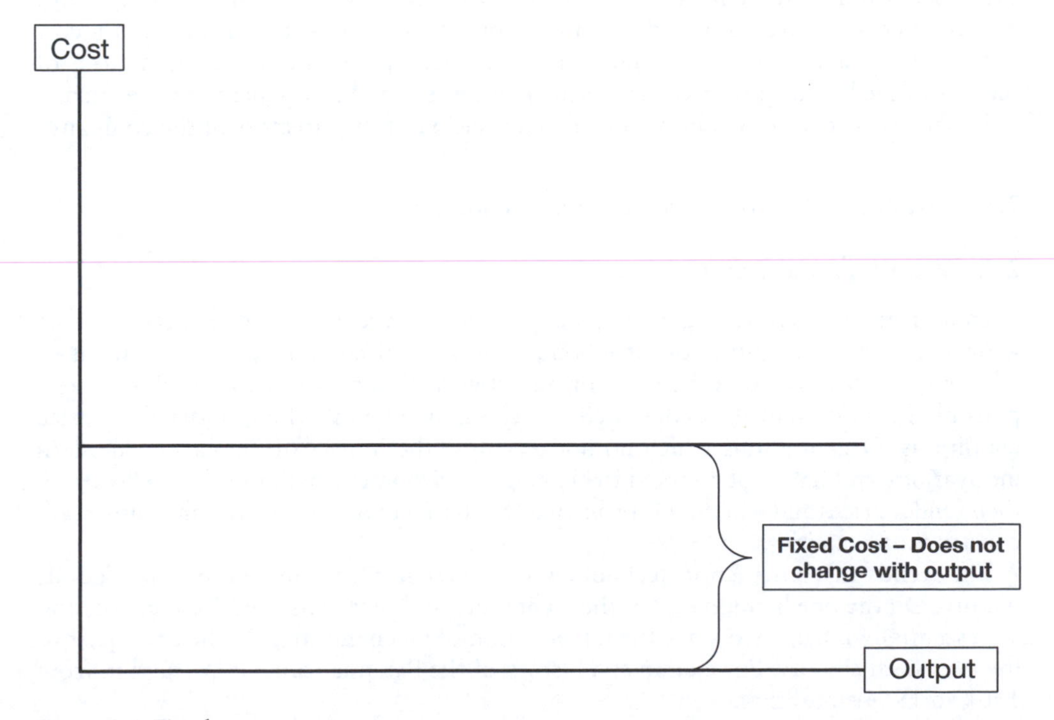

Figure 2.1 Fixed costs

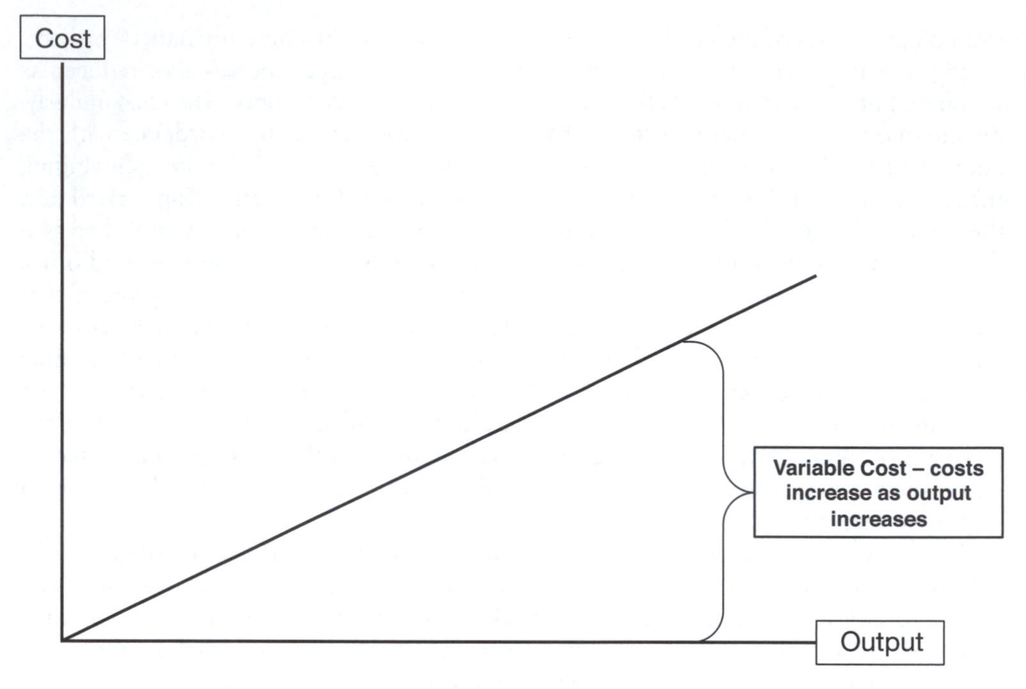

Figure 2.2 Variable costs

will then be considered, followed by the semi-variable costs of preliminaries and other site-based overheads and finally the more specific indirect costs of head office overheads. The chapter is completed by a discussion of profit and risk costs, which is not a cost in the true sense of the word, as no money is actually paid out. However, in terms of the title of the book, a profit and risk element is an essential component of the estimate, although the level of it will be a commercial decision of the directors of the company.

2.2 The influence of productivity on construction costs

2.2.1 What is productivity?

It should go without saying that good project management in construction must address the efficient utilisation of labour, material and equipment. Improvement of labour productivity should be a major and continual concern of those who are responsible for cost control of construction, especially when working under fixed price conditions. Organisations which do not recognise the impact of the various modern innovations and have not adapted to changing environments will find this reflected in their tender prices and will therefore find it difficult to remain competitive in mainstream construction activities.

The trends in construction technology can present a very mixed and ambiguous picture. On the one hand, many of the techniques and materials used for construction are essentially unchanged since the introduction of mechanisation in the early part of the twentieth century. For example, a history of the Panama Canal construction from 1904 to 1914 states that:

> The work could not have been done any faster or more efficiently in our day, despite all technological and mechanical advances in the time since, the reason being that no present system could possibly carry the spoil away any faster or more efficiently than the system employed. No motor trucks were used in the digging of the canal; everything ran on rails. And because of the mud and rain, no other method would have worked half so well.
>
> (McCullough, David, *The Path Between the Seas*, Simon and Schuster, 1977: 531)

In contrast, we can point to the continual change and improvements in traditional materials and techniques, even though on the face of it, the technology may not have appeared to change. For example, bricklaying has been carried out in much the same way for thousands of years; perhaps in the literal sense of placing brick on brick it has not changed very much. However, masonry technology has changed a great deal with motorised site transport replacing the hod carrier and modern chemically analysed mortars giving both stronger adhesion between bricks and additives/admixtures providing better weather protection and workability. Technological change is certainly occurring in construction, although often at a slower rate than in other sectors of the economy.

The construction industry often points to factors which cannot be controlled by the industry as an explanation of cost increases and the lack of technical innovation. These include HSE requirements, building regulations and labour agreements in an industry where trade unions have historically been strong. However, the construction industry should bear a large share of responsibility for not realising earlier that a technological edge which increases efficiency and productivity does have a direct effect on the bottom line (i.e. profit). The specious argument is usually dragged out that the industry cannot afford to invest in research and development because all projects are won by intense cost-based competition; however, the industry does spend considerable sums on legal fees related to the various disputes it generates. Money spent on development is an investment and therefore should generate additional revenue or savings in the future; money spent on legal fees does not have the same effect.

Individual design and/or construction firms should therefore explore new ways to improve productivity for the future. Of course, operational planning for individual construction projects is still important, but such short-term planning has limitations and may soon reach the point of diminishing returns because existing practices can only be made more efficient up to a point. What is needed now is strategic planning to usher in a revolution which can improve productivity by an order of magnitude. This strategic planning should look at opportunities and ask whether there are potential options which will give better outputs by the use of existing resources.

2.2.2 Productivity in the construction industry

Productivity on project sites is influenced by many factors which can be split into three distinct groupings:

a) Labour characteristics:
   - age, skill and experience of workforce
   - leadership and motivation of workforce.

b) Specific project conditions:
   - job size and complexity
   - job-site accessibility
   - labour availability
   - equipment utilisation
   - contractual agreements
   - local climate and weather conditions
   - local cultural characteristics, particularly outside of the contractor's home country.
c) Non-productive activities:
   - indirect labour required to maintain the progress of the project
   - rework for correcting unsatisfactory work
   - temporary work stoppage due to inclement weather or material shortage
   - time off for other non-job activities (e.g. trade unions)
   - absentee time, including late start and early finishes
   - non-working statutory holidays
   - strikes and other industrial action.

The non-productive activities associated with a project may or may not be paid for by the client, but they nevertheless take up potential labour resources which may otherwise be directed to fulfilment of the project. Each category of factors affects the productive labour available to a project in addition to the on-site labour efficiency.

Productivity of construction labour is often broadly defined as output per man-hour. Since labour constitutes a large part of the construction cost and the quantity of labour hours in performing a task in construction is more susceptible to the influence of management than are materials or capital, this productivity measure is often referred to as *labour productivity*. However, it is important to note that labour productivity is a measure of the overall effectiveness of an operating system in utilising labour, equipment and capital to convert labour efforts into useful output, and is not just a measure of the capabilities of labour alone. For example, by investing in a piece of new equipment to perform certain tasks in construction, output may be increased for the same number of labour hours, thus resulting in higher labour productivity.

Construction output may be expressed in terms of functional units or financially. In the former case, labour productivity is associated with units of product per labour hour, such as cubic metres of concrete placed per hour (m^3/hour) or kilometres of highway paved per day (km/day). In the latter case, the labour productivity is identified with value of construction (in the currency units) per labour hour (i.e. £/hour). The value of construction in this regard is not measured by the benefit of constructed facilities, but by construction cost. Labour productivity measured in this way requires considerable care in interpretation; for example, where wage rates in construction have been declining, the value of construction put in place per hour of work will decline as a result, suggesting lower productivity.

Labour characteristics

Performance analysis is a common tool for assessing the quality of individuals or teams and their contribution to the project completion. Factors that might be evaluated in such a performance analysis may include:

- *quality of work* – produced or accomplished
- *quantity of work* – volume of acceptable work in a particular time period
- *job knowledge* – i.e. has the person demonstrated knowledge of requirements, methods, techniques and skills involved in doing the job and in applying these to potentially increasing productivity?
- *related work knowledge* – knowledge of effects of work upon other areas and knowledge of related areas which may have influence on assigned work
- *judgement* – soundness of conclusions, decisions and actions
- *initiative* – ability to take effective action without being told
- *resource utilisation* – ability to delineate project needs and locate, plan and effectively use all resources available
- *dependability* – reliability in assuming and carrying out commitments and obligations
- *analytical ability* – effectiveness in thinking through a problem and reaching sound conclusions
- *communicative ability* – effectiveness in using oral and written communications and in keeping others adequately informed
- *interpersonal skills* – effectiveness in relating in an appropriate and productive manner to others
- *ability to work under pressure* – ability to meet tight deadlines and adapt to changes
- *security sensitivity* – ability to handle confidential information appropriately and to exercise care in safeguarding sensitive information
- *safety consciousness* – has knowledge of good safety practices and demonstrates awareness of own personal safety and the safety of others
- *profit and cost sensitivity* – ability to seek out, generate and implement profit-making ideas
- *planning effectiveness* – ability to anticipate needs, forecast conditions, set goals and standards, plan and schedule work and measure results
- *leadership* – ability to develop in others the willingness and desire to work towards common objectives
- *delegating* – effectiveness in delegating work appropriately
- *development people* – ability to select, train and appraise personnel, set standards of performance, and provide motivation to grow in their capacity.

These different factors could each be assessed on a three point scale: (1) recognised strength, (2) meets expectations, (3) area needing improvement. Examples of work performance in these areas might also be provided.

Both contractors and clients will be concerned with the labour activity and productivity at project sites, the contractor to ensure that the labour is working as effectively as possible and the clients because they have a health and safety responsibility. For this purpose, labour productivity is often expressed as functional units per labour hour for each type of construction task. However, even for such specific purposes, different levels of measure may be necessary. For example, cubic metres of concrete placed per hour is a more detailed measure than kilometres of highway paved per hour. More detailed measures are more useful for monitoring individual activities and personal performance, while higher-level measures may be more convenient for developing industry-wide standards of performance for organisations.

While each contractor or owner is clearly free to use their own system to measure labour productivity at a particular site, it is a good practice for the company to set up a system which can be used to track trends in productivity over time and in various locations. Considerable effort is required to collect information regionally or nationally over a number of years to produce such results, and the results are therefore jealously guarded by the company. The productivity indices compiled from statistical data should include parameters such as the performance of the major trades, the effects of project size, type of construction and location, together with other major project influences which are important to the company.

In order to develop industry-wide standards of performance which are published in the public domain in the standard pricing books, there must be a general agreement on the measures to be useful for compiling data. Then, the job-site productivity data collected by various contractors and owners can be correlated and analysed to develop measures for each of the major segments of the construction industry. In this way, a contractor or client can compare the performance of its project with that of the industry average. This process is known as benchmarking.

Specific project conditions

Project site labour productivity can be estimated either for each trade (carpenter, bricklayer, etc.) or each type of construction (residential housing, processing plant, etc.) under a specific set of work or site conditions. A *base labour productivity* may then be defined for a set of work conditions specified by the client or contractor who wishes to observe and measure the labour performance over a period of time under such conditions. A *labour productivity index* may then be defined as the ratio of the project labour productivity under a different set of work conditions to the base labour productivity, and is a measure of the relative labour efficiency of a project under this new set of work conditions.

The effects of various factors related to work conditions on a new project can be estimated in advance, some more accurately than others. For example, for very large construction projects, the labour productivity index tends to decrease as the project size and/or complexity increases because of logistic problems and the learning curve that the workforce must undergo before adjusting to the new environment. Site accessibility may also reduce the labour productivity index if the workers are constrained in how they can perform their jobs, such as working on sites that are still being used by the client. Labour availability in the local market may be another factor as shortage of local labour will force the contractor to bring in non-local labour or schedule overtime work or both. In either case, the labour efficiency will be reduced in addition to incurring additional expenses. The degree of equipment utilisation on a construction project will clearly have a further direct bearing on site labour productivity, as will the use of subcontractors and the degree of field supervision, all of which will impact job-site labour productivity. Since on-site construction essentially involves outdoor activities in the first half of the project, the local climate will also directly influence the efficiency of workers. When working outside the contractor's own country, the cultural characteristics of the host country should be observed in assessing the labour efficiency, which can have a major effect, e.g. reduced working hours during the holy month of Ramadan for Muslim workers.

Non-productive activities

The non-productive activities associated with a project should also be examined in order to assess the *productive labour yield*, which is defined as the ratio of direct labour hours devoted to the completion of a project to the potential labour hours. The direct labour hours are estimated on the basis of the best possible conditions at a project site by excluding all factors which may reduce the productive labour yield. For example, in the repaving of highway surface, the flagmen required to divert traffic represent indirect labour which does not contribute to the labour efficiency of the paving crew. Similarly, for large projects in remote areas, indirect labour may be used to provide housing and infrastructure for the workers hired to supply the direct labour for a project. The labour hours spent on rework to correct unsatisfactory or rejected original work represent extra time taken away from potential labour hours. The labour hours related to such activities must be deducted from the potential labour hours in order to obtain the actual productive labour yield.

2.2.3 Construction processes

The previous sections described the primary inputs of labour, material and equipment to the construction process. At varying levels of detail, a project manager must ensure that these inputs are effectively coordinated to achieve an efficient construction process. This coordination involves both strategic decisions and tactical management in the field. For example, strategic decisions about appropriate technologies or site layout are often made during the process of construction planning. During the course of construction, foremen and site managers will make decisions about work to be undertaken at particular times of the day based upon the availability of the necessary resources of labour, materials and equipment. Without coordination among these necessary inputs, the construction process will be inefficient or stop altogether.

Example: steel erection

Erection of structural steel for buildings, bridges or other facilities is an example of a construction process requiring considerable coordination. Fabricated steel pieces must arrive on site in the correct order and quantity for the planned effort during a day. Crews of steelworkers must be available to fit pieces together, bolt joints, and perform any necessary welding. Cranes and crane operators may be required to lift fabricated components into place; other activities on a job site may also be competing for use of particular cranes. Welding equipment, wrenches and other hand tools must be readily available. Finally, ancillary materials such as bolts of the correct size must be provided.

In coordinating a process such as steel erection, it is common to assign different tasks to specific crews. For example, one crew may place members in place and insert a few bolts in joints in a specific area. A following crew would be assigned to finish bolting, and a third crew might perform necessary welds or attachment of brackets for items such as curtain walls.

With the required coordination among these resources, it is easy to see how poor management or other problems can result in considerable inefficiency. For example, if a shipment of fabricated steel is improperly prepared, the crews and equipment on site may have to wait for new deliveries.

Queues and resource bottlenecks

A project manager needs to ensure that resources required for and/or shared by numerous activities are adequate. Problems in this area can be indicated in part by the existence of queues of resource demands during construction operations. A *queue* can be a *waiting line* for service. One can imagine a queue as an orderly line of customers waiting for a stationary server such as a ticket seller. However, the demands for service might not be so neatly arranged. For example, we can speak of the *queue* of welds on a building site waiting for inspection. In this case, demands do not come to the server, but a roving inspector travels among the waiting service points. Waiting for resources such as a particular piece of equipment or a particular individual is an endemic problem on construction sites. If workers spend appreciable portions of time waiting for particular tools, materials or an inspector, costs increase and productivity declines. Insuring adequate resources to serve expected demands is an important problem during construction planning and field management.

In general, there is a trade-off between waiting times and utilisation of resources. Utilisation is the proportion of time a particular resource is in productive use. Higher amounts of resource utilisation will be beneficial as long as it does not impose undue costs on the entire operation. For example, a welding inspector might have 100 per cent utilisation, but workers throughout the job site might be wasting inordinate time waiting for inspections. Providing additional inspectors may be cost effective, even if they are not utilised at all times.

2.3 Labour costs

Employing people involves many additional costs to a company over and above the person's actual wage or salary. These are known as on-costs and must be included when calculating the costs of carrying out the project. The on-costs for the various construction trades are covered in the Working Rule Agreement (WRA), prepared by the Construction Industry Joint Council (CIJC) and include the following costs which an employer is obliged to cover:

- *Guaranteed minimum wage* appropriate for the particular employee.
- *Contractor's bonus allowance*, which will depend on the employee's performance and productivity.
- *Inclement weather allowance*, as the contractor will incur costs without any productivity.
- *Non-productive overtime*, employees may be paid 'time-and-a-half' for overtime, therefore the contractor will be required to pay 1.5 hours wage for 1 hour productivity, therefore the extra half hour is classed as non-productive overtime. The National Working Rule (NWR) 7 gives the rules for calculating overtime rates.
- *Sick pay allowance*, another cost which is not covered by productivity. NWR 16 covers this allowance.
- *Trade supervision*, a cost which may or may not be included by contractors and covers non-productive time by the leader of a group of tradesmen or labourers (sometimes referred to as 'gangers').
- *Working rule agreement allowances*, which includes various extra payments for labourers engaged in particular activities, for example working in difficult

conditions, shift and night-time working or continuous working over meal and break times. NWR 3 and 4 includes the full list of activities covered.
- *CITB training contributions*, which are levied at set percentages for companies registered with the Construction Industry Training Board. The contributions are at different rates for employed staff and self-employed and labour-only subcontractors.
- *National Insurance contributions*, the employer's contributions are included here as the employee's contributions would be part of their wage or salary.
- *Holiday credits*, payments for statutory and annual holidays.
- *Tool allowance*, for particular trades where WRA 18 allows payment to tradesmen using their own tools.
- *Severance payment*, in accordance with statutory allowances.
- *Employer liability insurance*, although this may be included as a project overhead in the preliminary section of the project costs.

Other labour costs which would normally be included in the project overheads include:

- *Daily travel allowance*, depending on whether the employer provides transport.
- *Periodic leave and lodging allowance*, if staff are required to stay overnight close to the project.
- *Supervision*, this is a non-productive cost and is usually included as a preliminary item in a bill of quantities.
- *Attraction money*, which may be required in remote areas where there is a shortage of local labour.

It may not be possible, or economic, to determine all of these payments for each individual project especially where the total labour requirement cannot be accurately ascertained. Some of the above costs may be covered in the project overheads rather than direct costs and this will be a commercial decision of the individual company. However, the contractor will be exposed to all of the above costs and allowance should be made for them somewhere in the total project estimate.

The contractor will also be very conscious that projects are tendered in competition with other firms, who may have a different cost structure. Therefore, care must be taken when purely building up the costs from first principles that the company's costs do not become uneconomic, meaning that they may never win a tender in competition. This is a very difficult call to make by the company's directors and requires a great deal of experience and market knowledge, which will be discussed further in Chapter 7.

2.3.1 All-in rate for labour

Because of the difficulty of calculating direct labour costs on a project-by-project basis, most companies calculate these costs on an annual basis for the various categories of direct labour that they employ. This also clearly reduces the burden on the project estimating function. These direct labour costs are calculated in a three-stage process:

1. Establish the number of working hours in a one-year period.
2. Establish the annual cost to the company of the items above. The total will be the annual cost of employing an operative in each of the chosen categories.
3. Divide (2) by (1) to establish the all-in hourly rate of employing the operative.

If the duration of the project is anticipated to take more than one year, it would, however, be worth calculating this rate as specific for the project, especially if the contract terms and conditions do not allow increased costs (i.e. inflation) to be paid to the contractor.

See section 6.6.1 for a worked example of an all-in rate for labour.

2.3.2 Number of hours worked

In terms of calculating the number of hours worked during a calendar year, because construction projects require people to work outside (at least until the envelope has been completed), the working hours will generally be different in summer and winter. The actual hours worked by operatives will be different for different companies and some regional variation is also to be expected. The WRA in the UK assumes an annual average of a 39-hour week based on 5 days of 8 hours per day with one hour less on Fridays. If the local custom is different, this will clearly affect the calculation of the all-in rate for labour.

For example, in hot climates such as the Middle East, working hours will be shorter in the summer than the winter, although night work in the summer months is quite common in order to perform certain operations such as concrete pouring as the concrete would dry too quickly in the daytime heat. In addition, working hours during the month of Ramadan will be reduced for Muslim workers. All these local issues will need to be taken into account during the estimating period.

The labour costs of performing a defined amount of work will depend, not only on the cost of the manpower but also on the productivity, or efficiency, with which the labour can complete the work. Most construction firms will keep their own records of labour outputs and they are also available in public domain pricing books, such as Spon's. Care must be taken, however, in using standardised information from either previous projects or pricing books because each project will have particular site conditions which need to be considered and also, due to the time required for publishing, the data may well be out of date, especially if there has been a considerable increase or decrease in construction costs or tender prices.

2.4 Materials costs

Materials represent a major expense in construction and, theoretically, all contractors who tender for a project should have the same materials costs, since many of the materials come from the same source anyway. Therefore, minimising procurement or purchase costs presents important opportunities for reducing overall costs and creating a more competitive bid price. Poor materials management during the construction stage can also result in large and mainly avoidable unproductive costs during construction on site. If materials are purchased early, the contractor may benefit from lower costs, but their capital may be tied up for longer periods and the company may also have increased costs of security, protection, insurance etc. The contractor needs to know how long in advance of delivery they need to order the materials (called the lead time) as it is obviously essential that the materials are available on site at the time they are programmed to be installed, otherwise additional and unnecessary costs will be incurred. See also section 2.4.6 on just in time delivery. Some forms of contract allow the contractor to be paid for materials stored on site or even materials stored off site, but this practice is becoming increasingly rare and the costs associated with keeping stock ('inventories' in US terminology) is for the contractor to absorb.

Materials management is not just a concern during the construction stage. Decisions about materials procurement will also be required during the initial planning and scheduling stages. For example, activities can be inserted in the project schedule to represent procurement of major items, especially those relating to mechanical and electrical installations which require substantial design and factory manufacture. The availability of such materials and systems may greatly influence the schedule in projects with a very tight time schedule and sufficient time for obtaining the necessary products must be allowed for.

Materials management may also create difficulties at the organisation level if central purchasing and inventory control (by the contractor) is used for standard items. In this case, the various projects undertaken by the organisation would present requests to the central purchasing group, which would maintain inventories of standard items to reduce the delay in providing material or to obtain lower costs due to bulk purchasing. This organisational materials management problem is analogous to inventory control in any organisation facing continuing demand for particular items.

Materials ordering lends itself particularly well to computer-based systems in order to ensure the consistency and completeness of the purchasing process. In the manufacturing industry, the use of automated *materials requirements planning* systems is common, where the master production schedule, inventory records and product component lists are merged to determine what items must be ordered, when and how much is required. The heart of these calculations is simple; the projected demand for each material item in each period is subtracted from the available stock in the company's stores. When the stock becomes too low, a new order is placed. For items that are non-standard or not kept in stock, the calculation is even simpler since no inventory must be considered. With a materials requirement system, much of the detailed record keeping is automated and project managers are alerted to purchasing requirements. However, this is all very well in theory but often contractors do not know when they are going to win jobs and permanent stock can only be kept for very standard items like bricks and cement, which are usually immediately available from suppliers anyway.

2.4.1 Who owns the materials?

Under many standard forms of contract, payment is made to the contractor for 'work done and materials reasonably and properly brought to site' and so work is paid for after the event, rather than in advance. The contractor is therefore not paid for the materials until they are physically on site and in a lot of cases, physically incorporated into the works. In such circumstances, a contractor would be well advised to understand who actually own the materials in order to protect their position.

Most standard forms of contract provide that title (i.e. ownership) of the materials passes to the Employer when materials are included in a payment certificate issued to the main contractor, rather than upon delivery to site. This means that the contractor owns the materials whilst they are stored on site but the employer owns them once they have been valued and certified as part of the project, even though the certificate may not have been paid yet. This risk is compounded for subcontractors, whose materials may have been included in a payment certificate between the main contractor and employer before the main contractor has paid the subcontractor for them. Subcontractors and suppliers should therefore take note, even if the title to materials has not passed to the main contractor, the employer can still obtain a superior title to the materials. Further discussions of legal points such as this are not part of the remit of this book but as it is possible that more than one party can have title to the same materials, it is a question of ensuring your title claim takes priority over any other.

Retention of title clauses

A well-drafted retention of title clause is important for anyone supplying building materials. It will have the effect of retaining the supplier's title to any unfixed materials. If the customer is a main contractor and has received a payment certificate from the employer covering the value of the materials, then the retention of title clause should give an unpaid subcontractor's title priority over the employer's, as long as the employer is notified of the clause prior to delivery.

Fixed materials

It is more difficult to retain title to materials that have been built in to the project works. Ownership of materials passes to the employer (and/or to any landowner) as soon as materials are fixed, regardless of payment. There have been a number of cases where an unscrupulous property owner with knowledge of this law has refused to pay a builder for works by dreaming up some spurious excuses, and the builder cannot reclaim any of the materials or equipment that have been built in to the property. Express agreement with the employer or landowner is required to change the position. Subcontract retention of title clauses will assist a contractor only if brought to the employer and/or landowner's attention before the materials are fixed into the project.

Part-fixed materials

What happens if the materials are part installed when the retention of title dispute arises? To establish at what point the materials became fixtures, the relevant factors are:

- the intentions of the installers, i.e. has the fixing operation already started, or are the materials merely piled up at convenient locations ready to be fixed
- the degree of annexation, i.e. the extent to which the goods have become incorporated within or fixed to the permanent structure
- to what extent the precise steps needed to install the materials in question have been carried out and
- the steps required to remove the materials in a condition to be resold.

The overriding principle is whether the fixing of each item in question was intended to enable it to be used

a) as an item in its own right, or
b) to enable the land or building to be more conveniently used and enjoyed as a piece of real property.

If (a) applies, title may not have passed, but if (b) applies, title has passed unless the parties previously agreed otherwise.

If title has passed, as mentioned above, an unpaid contractor is in a no-win situation. On the one hand, they are not entitled to remove the materials even if the employer disconnects them. On the other, and to add insult to injury, they may be required to honour any ongoing maintenance obligation, where they may be required to remove and replace defective materials.

2.4.2 Discounts

In older standard forms of contract (e.g. the 1980 edition of JCT), main contractor's discount, or MCD, related to a cash discount offered by the supplier or subcontractor to a main contractor in return for prompt payment. It generally ranged from 2.5 per cent for subcontracts to 5 per cent for the supply only of materials.

One function of the Construction Act was to try and get rid of the historical problems suppliers and contractors experienced with payment and as such MCDs became less common (there is no mention of them in the 1998 or 2005 suites of JCT contracts). However, cash discounts and the like are still mentioned in documents such as the National Schedule of Rates and the CIOB's Code of Estimating Practice.

2.4.3 Wastage

Any form of waste is an unnecessary cost to the contractor. The client normally only pays the contractor for work which is properly carried out under the scope of works of the project, so anything which is not part of the scope of works will not be paid for. The exception to this would be projects where the contractor is paid on a cost-plus or reimbursable basis. In this case, the contractor would be paid for resources which are brought onto the site, which may include those resources which are not used (i.e. wastage). Unless the contractor is lucky enough to be engaged on a cost reimbursable project, they would normally be well advised to minimise waste as much as possible. Additionally, the modern trend of environmental awareness, which is often backed up by legislation, has meant that all contractors have a legal as well as moral duty to minimise the waste that they produce.

What is waste?

Waste is actually quite a complex area. Superficially, waste is any substance or object(s) that the company discards or intends to discard and therefore includes any material which is perceived to have no further use. Waste can be classed as:

- inert (e.g. bricks, glass)
- hazardous (e.g. paint tins, mastic tubes)
- non-hazardous (e.g. packaging, plasterboard).

However, this only relates to waste products and takes no account of wastage or inefficiencies in construction processes, which can be just as costly to the contractor but has little or no physical effect.

In terms of physical waste, all businesses have a duty to ensure that waste they produce is handled safely and within the law. The contractor is legally obliged to check that any company removing waste from the site is a registered waste carrier, and that they take the waste to a registered waste management site. Every load of physical waste that leaves the site must be accompanied by a waste transfer note, which provides an accurate description of the waste to enable it to be treated safely at its destination. Under EU law, the originator of the waste (i.e. contractor) is required to keep copies of all waste transfer notes for at least two years. A site waste management plan (SWMP) is a legal requirement in England for projects over £300,000. It requires the contractor to forecast and record waste and how it is managed, thereby at least giving some thought to how waste can be minimised.

Apart from the obvious source of waste, such as packaging, used paint tins, etc., waste can also be generated in other ways:

- *Site management* – inefficient and untidy sites will generate more waste than is necessary as materials will be damaged more easily if not stored properly and would also suffer more losses and theft if there is poor site control.
- *Manufacture* – products which are manufactured in non-standard sizes, requiring cutting on site, or poor product information supplied by the manufacturer may increase the wastage element in the project.
- *Supply and logistics* – poor handling and transportation by the supplier may create damage or loss which inevitably increases wastage. Storage, too early or too late delivery and inappropriate delivery methods may add unnecessary costs.
- *Design wastage* – issues such as time constraints, product awareness, design coordination and product compatibility at the design stage may create additional wastage costs during the construction stage.

Wastage can therefore add significant extra costs to the contractor, some of which are clear and obvious, while others are hidden and are more concerned with inefficient and uneconomic processes both in the design stage and on the construction site. As we have mentioned elsewhere in this chapter, costs are incurred on site during the construction stage but these costs are often generated in the design stage.

2.4.4 Material procurement and delivery

The main sources of information for feedback and control of material procurement are requisitions, bids and quotations, purchase orders and subcontracts, shipping and receiving documents, and invoices. For projects involving the large-scale use of critical resources, the owner may initiate the procurement procedure even before the selection of a constructor in order to avoid shortages and delays. Under ordinary circumstances, the constructor will handle the procurement to shop for materials with the best price/performance characteristics specified by the designer. Some overlapping and multiple handling in the procurement process is unavoidable, but it should be minimised to ensure timely delivery of the materials in good condition.

The materials for delivery to and from a construction site may be broadly classified as:

1 bulk materials,
2 standard off-the-shelf materials, and
3 fabricated members or units.

The process of delivery, including transportation, field storage and installation will be different for each of these classes of materials.

Bulk materials generally refer to materials in their natural or semi-processed state, such as earthwork to be excavated, wet concrete mix, etc. which are usually encountered in large quantities in construction. Some bulk materials such as earthwork or gravels may be measured in bank (solid in situ) volume. Obviously, the quantities of materials for delivery may be substantially different when expressed in different measures of volume, depending on the characteristics of such materials.

Standard piping and valves are typical examples of standard off-the-shelf materials which are used extensively in the chemical processing industry. Since standard off-the-shelf materials can easily be stockpiled, the delivery process is relatively simple.

Fabricated members such as steel beams and columns for buildings are pre-processed in a shop to simplify the field erection procedures. Welded or bolted connections are attached partially to the members which are cut to precise dimensions for adequate fit. Similarly, steel tanks and pressure vessels are often partly or fully fabricated before shipping to the field. In general, if the work can be done in the shop where working conditions can better be controlled, it is advisable to do so, provided that the fabricated members or units can be shipped to the construction site in a satisfactory manner at a reasonable cost.

As a further step to simplify field assembly, an entire wall panel including plumbing and wiring or even an entire room may be prefabricated and shipped to the site. While the field labour is greatly reduced in such cases, 'materials' for delivery are in fact manufactured products with value added by another type of labour. With modern means of transporting construction materials and fabricated units, the percentages of costs on direct labour and materials for a project may change if more prefabricated units are introduced in the construction process.

In the construction industry, materials used by a specific craft are generally handled by craftsmen, not by general labour. Thus, electricians handle electrical materials, pipefitters handle pipe materials, etc. This multiple handling diverts scarce skilled craftsmen and contractor supervision into activities which do not directly contribute to construction. Since contractors are not normally in the freight business, they do not perform the tasks of freight delivery efficiently. All these factors tend to exacerbate the problems of freight delivery for very large projects.

2.4.5 Stock control

Once goods are purchased, they become part of the working capital of the company. if you look at any published balance sheet for a company, the working capital (part of the company's current assets) consist of:

a) cash held in the company's bank accounts, *which is used to pay for [. . .]*
b) stock (inventories), *which include the materials to be built into the project, thus creating [. . .]*
c) work in progress, *as the materials are being incorporated – when completed, they are [. . .]*
d) finished goods, *not yet paid for, so the client joins the [. . .]*
e) debtors, *i.e. clients who owe money to the company – when paid, this will* increase *cash in (a)*
f) creditors *(when materials are bought on credit terms, these are the suppliers etc. to whom the company owes money. When paid, this will* reduce *cash in (a))*
g) and the merry-go-round starts again. This is known as the *working capital cycle.*

The general objective of stock control is to minimise the cash tied up in this working capital cycle by making trade-offs among the major categories of costs: (1) purchase costs, (2) order cost, (3) storage costs and (4) unavailable cost. These cost categories are interrelated since reducing cost in one category may increase cost in others. The costs in all categories generally are subject to considerable uncertainty.

2.4.6 Just in time

Just in time (JIT) is a production and stock control system in which materials and components are delivered as close as possible to the point that they are incorporated into the works. Stocks held on site are therefore reduced to the minimum and in some cases to zero, if the materials are delivered direct to the work location (as often happens in some manufacturing industries). Traditionally, contractors have maintained stock levels both on and off site to act as buffers so that operations can proceed smoothly even if there are unanticipated disruptions to supply. While these stock levels provide buffers against unforeseen events, they clearly have a cost, as stated below. In addition to the funds tied up in the stock themselves, the absence of stock, or rather the lack of an immediate replacement may encourage the operative to get their work right first time, resulting in fewer defects and wastage.

Under ideal conditions, a company operating a JIT supply system would purchase only enough materials each day to meet that day's needs. Moreover, the company would have no raw materials in stock at the end of the day and all work completed during the day is consequently part of the finished works. As this sequence suggests, 'just in time' means that raw materials and components are received just in time to be incorporated into the works.

Although few contractors have been able to reach this ideal for mainly practical reasons, many contractors have been able to reduce stock levels to a fraction of their previous levels. The result has been a substantial reduction in ordering and warehousing costs, and much more efficient and effective site operations. In a just in time environment, the flow of goods is controlled by a 'pull' approach, where the stock is only delivered following a downstream requisition or confirmed need. At the final construction stage a signal is sent regarding the exact amount of parts and materials that would be needed over the next few hours to complete the work scheduled during that period and only that amount of materials and components are provided. The same signal is sent back to each preceding stage in the supply chain so that a smooth flow of components and materials is maintained with no appreciable stock build-up at any point. Thus all stages of the supply chain respond to the pull exerted by the final construction stage, which is directed by the construction schedule. Under a just in time system you don't produce anything, anywhere, for anybody unless they ask for it somewhere downstream.

2.4.7 Types of material costs

Purchase costs

The purchase cost of an item is the unit purchase price from an external source including transportation to site. For construction materials, it is common to receive discounts for bulk purchases (economies of scale), so the unit purchase cost should go down as the quantity ordered increases. These reductions may reflect manufacturers' marketing policies, economies of scale in the material production, or in transportation. There are also advantages in ordering large quantities to ensure, for example, the same colour or size of items such as facing bricks. Therefore, making a limited number of large purchases for materials has considerable economic benefits. Even for smaller companies, it may be possible to consolidate small orders from a number of different projects to benefit from these economies.

The cost of materials is often based on prices obtained through negotiation and bargaining. Unit prices of materials will invariably depend on bargaining leverage, quantities ordered and the required delivery time. Organisations with potential for long-term purchase volume will be able to command better bargaining leverage. While large orders may result in lower unit prices, they may also increase storage costs and thereby affect the contractor's cash flow in a different way. Requirements of short delivery time can also adversely affect unit prices. Furthermore, design characteristics which include items of odd sizes or shapes should be avoided if possible. Since such items normally are not available in the standard stockpile, purchasing them causes higher prices. The transportation costs are also affected by shipment sizes as shipment by the full load often reduces prices and assures speedier delivery.

Order cost

This reflects the administrative expense of issuing a purchase order to an outside supplier. Order costs include expenses of making requisitions, analysing alternative vendors, writing purchase orders, receiving materials, inspecting materials, checking on orders and maintaining records of the entire process. Order costs are usually only a small proportion of total costs for material management in construction projects, although ordering may require substantial time.

Storage costs

These costs are primarily the costs of handling, storage, obsolescence, shrinkage and deterioration. Capital cost results from the opportunity cost or financial expense of capital tied up in stock, although this is not strictly a physical cost but an opportunity cost. Once payment for goods is made, borrowing costs are incurred or capital must be diverted from other uses. Consequently, a capital carrying cost is incurred equal to the value of the inventory during a period multiplied by the interest rate obtainable or paid during that period. Note that capital costs only accumulate when payment for materials actually occurs; many organisations attempt to delay payments as long as possible to minimise such costs. Handling and storage represent the movement and protection charges incurred for materials. Storage costs also include the disruption caused to other project activities by large inventories of materials that get in the way. Obsolescence is the risk that an item will lose value because of changes in specifications. Shrinkage is the decrease in inventory over time due to breakage, theft or loss. Deterioration reflects a change in material quality due to age or environmental degradation. Many of these storage cost components are difficult to predict in advance; a project manager knows only that there is some chance that specific categories of cost will occur. In addition to these major categories of cost, there may be ancillary costs of additional insurance, taxes, or additional fire hazards. As a general rule, holding costs will typically represent 20 to 40 per cent of the average stock value over the course of a year.

Unavailability cost

This cost is incurred when a desired material is not available at the desired time. Shortages may delay work, thereby wasting labour resources or delaying the completion of the entire project. Again, it may be difficult to forecast in advance exactly when an item may be required or when a shipment will be received. While the project schedule gives one estimate, deviations from the schedule may occur during construction. Moreover, the cost associated with a shortage may also be difficult to assess; if the material used for one activity is not available, it may be possible to assign workers to other activities and, depending upon which activities are critical, the project may not be delayed.

2.5 Plant and equipment costs

The selection of the appropriate type and size of construction plant and equipment often affects the required amount of time and effort and thus the job-site productivity of a project. It is therefore important for both construction planners and estimators to be familiar with the characteristics of the major types of equipment most commonly used on construction projects.

The introduction of new mechanised equipment in construction has had a profound effect on the cost and productivity of construction as well as the methods used for construction itself. A modern example of innovation in this regard is the introduction of microchips on tools and equipment, which allows the performance and activity of equipment to be continually monitored and adjusted for improvement. In many cases, automation of at least part of the construction process is possible and desirable. For example, wrenches that automatically monitor the elongation of bolts and the applied torque can be programmed to achieve the best bolt tightness. On civil engineering projects, laser-controlled scrapers can produce more accurate cuts faster and more precisely than methods which rely solely on the operator's skill and experience.

2.5.1 Choice of equipment and production outputs

Typically, construction equipment is used to perform essentially repetitive operations, and can be broadly classified according to two basic functions:

1 operators such as cranes, graders, etc. which stay within the confines of the construction site
2 haulers such as lorries, dumper trucks, ready mixed concrete trucks etc. which transport materials to and from the site.

In both cases, the cycle of a piece of equipment is a sequence of tasks which is repeated to produce a unit of output. For example, the sequence of tasks for a crane might be to fit and install a wall panel (or a package of wall panels) on the side of a building; similarly, the sequence of tasks of a ready mixed concrete truck might be to load, haul and unload fresh concrete during a programmed concrete pour.

In order to increase job-site productivity, the contractor should always select the most suitable equipment for the work conditions at a particular construction site. For example, in the excavation stage, factors which would affect the selection of excavators include:

1 *Size of the job.* Larger volumes of excavation will require larger excavators, or smaller excavators in greater number.
2 *Activity time constraints.* Shortage of time for excavation may force contractors to increase the size or numbers of equipment for activities related to excavation.
3 *Availability of equipment.* Productivity of excavation activities will diminish if the equipment used to perform them is available but not the most adequate.
4 *Cost of transportation of equipment.* This cost depends on the size of the job, the distance of transportation and the means of transportation.
5 *Type of excavation.* Principal types of excavation in building projects are cut and/or fill, excavation massive, and excavation for the elements of foundation. The most adequate equipment to perform one of these activities is not the most adequate to perform the others.
6 *Soil characteristics.* The type and condition of the soil is important when choosing the most adequate equipment since each piece of equipment has different outputs for different soils. Moreover, one excavation pit could have different soils at different strata.
7 *Geometric characteristics of elements to be excavated.* Functional characteristics of different types of equipment make such considerations necessary.
8 *Space constraints.* The performance of equipment is influenced by the spatial limitations for the movement of excavators.
9 *Characteristics of haul units.* The size of an excavator will depend on the haul units if there is a constraint on the size and/or number of these units.
10 *Location of dumping areas.* The distance between the construction site and dumping areas could be relevant not only for selecting the type and number of haulers, but also the type of excavators.
11 *Weather and temperature.* Rain, snow and severe temperature conditions affect the job-site productivity of labour and equipment.

By comparing various types of machines for excavation, for example, power shovels are generally found to be the most suitable for excavating from a level surface and for attacking an existing digging surface or one created by the power shovel; furthermore, they have the capability of placing the excavated material directly onto the haulers. Another alternative is to use bulldozers for excavation.

The choice of the type and size of haulers is based on the consideration that the number of haulers selected must be capable of disposing of the excavated materials expeditiously. Factors which affect this selection include:

1 *Output of excavators*: the size and characteristics of the excavators selected will determine the output volume excavated per day.
2 *Distance to dump site*: sometimes part of the excavated materials may be piled up in a corner at the job site for use as backfill.
3 *Probable average speed*: the average speed of the haulers to and from the dumping site will determine the cycle time for each hauling trip.

4 *Volume of excavated materials*: the volume of excavated materials including the part to be piled up should be hauled away as soon as possible.
5 *Spatial and weight constraints*: the size and weight of the haulers must be feasible at the job site and over the route from the construction site to the dumping area.

Dumper trucks are usually used as haulers for excavated materials as they can move freely with relatively high speeds on city streets as well as on highways, although they will need to be publicly licensed to be able to do this.

The cycle capacity of a piece of equipment is defined as the number of output units per cycle of operation under standard work conditions. The capacity is a function of the output units used in the measurement as well as the size of the equipment and the material to be processed. The cycle time refers to units of time per cycle of operation. The standard production rate of a piece of construction equipment is defined as the number of output units per unit time.

Each of these various adjustment factors must be determined from experience or observation of job sites. For example, a bulk composition factor is derived for bulk excavation in building construction because the standard production rate for general bulk excavation is reduced when an excavator is used to create a ramp to reach the bottom of the bulk and to open up a space in the bulk to accommodate the hauler.

2.5.2 Hire or buy?

Once the decision has been made regarding the actual equipment to be used, the contractor needs to make a further decision regarding whether to buy the equipment or to hire it as required from a specialist plant hire company. There are clearly major cost implications in this decision, since buying the equipment will involve a major capital outlay and also incur additional annual costs of storage, depreciation, maintenance etc. but it will be available for other projects and in the longer term will be a more economical solution providing that the equipment is used regularly. On the other hand, hiring the equipment will usually work out more expensive per hour, but does not involve an initial capital outlay and there are no annual maintenance charges. The CIOB Code of Estimating Practice gives the following list, which must be taken into consideration when purchasing items of plant:

a) *Purchase price (P) less expected resale value (R).* This is effectively a calculation of the depreciation. If the item of plant is expected to have a useful life of (x) years, then the depreciation (on a straight line basis) would be

$$\frac{((P) - (R))}{(x)}$$

Which is the annual capital cost of ownership of the equipment, and subject to tax rules and capital allowances, would be classed as a cost to the company in the profit and loss account. Note though that no actual money leaves the company's bank account due to this 'cost'.

b) *Return required on capital invested.* Once it has been bought, the piece of plant/equipment becomes an asset of the company and appears as such on the company's balance sheet. All assets are required to work for their living, so as long as the equipment is being occupied on a project, the return to the company is the costs it is not paying out for hiring the item of plant. The more it is used, the greater the return.

c) *Cost of finance*. Did the company have to borrow the money to buy the equipment? If not, what else would they have done with the money used to buy the equipment – opportunity cost.
d) *Direct costs of maintaining the equipment and overhead costs of the maintenance yard*. Costs of spares, including stock levels, and indirect costs of the maintenance facility.
e) *Direct costs of insurance, taxes, licences etc. required to operate the equipment.*
f) *Availability and cost of equivalent plant/equipment in the local market*. If the local market is particularly competitive, there may be little point in owning equipment outright, even if it is likely to be highly utilised.
g) *Proximity of company servicing centre to the project location*. Large items of plant may be very expensive to transport around the country, therefore, hiring locally with the hire company responsible for transporting to site may be beneficial.

2.6 Subcontractors' costs and attendances

2.6.1 Types of subcontractor

As recently as the 1970s, contractors executed a considerable amount of work themselves, subcontracting only that work which they were not geared up to do or which could be done cheaper by others. Common trades that were subcontracted included trades that either needed specialist tools and equipment, or specialist skills which the main contractor may not have need for on a full-time permanent basis. In the modern construction industry, it is common to find almost all trades being subcontracted, so that the modern main contractor has become more or less a manager of other subcontractors or 'work packages' as they are also known in some procurement methods.

Domestic subcontractors

Domestic subcontractors are usually selected entirely by the main contractor. Whilst most forms of contracts require the approval of the client's consultant or contract administrator, this is usually a formality and the contractor is given complete freedom to select their own subcontractors if no specific requirement to subcontract is given.

Named subcontractors

In most cases, named subcontractors are treated as domestic subcontractors. The contractor may be given a limited choice (a number of firms may be listed in the contract from which the contractor may select one of them to be a domestic subcontractor) or they may have no option but to use a single firm named in the contract. Clearly, the latter case can only occur in countries where there are no competition rules, or where there are protectionist rules covering local companies. Where there is only a single firm named, there is unlikely to be a highly competitive bid from the subcontractor. However, the use of named subcontractors does enable the designer to have some control over the selection of firms to carry out certain specialist work.

Nominated subcontractors (not appropriate for JCT05 and later editions)

As the term implies, these subcontractors are selected and nominated by the employer or their consultants. Apart from rules regarding valuation and payment for work done by nominated subcontractors and (in the case of JCT98 contracts) consent to grant extensions of time, the nominated subcontractors are very little different from domestic subcontractors.

The other major area of the main contractor's 'supply chain' is that of the suppliers of materials and plant. Both contractors and subcontractors rely on a wide range of suppliers for the goods and materials to be incorporated in the works and again, these are grouped into 'domestic' and 'named/nominated'.

Domestic suppliers

Unless the suppliers are nominated or specified, the contractor or subcontractor will be free to purchase materials from any supplier they choose provided that they comply with the specification. Approval of suppliers may be a condition of the contract but this will usually be a formality unless there are special provisions regarding source or sample etc.

Named or specified suppliers

Named suppliers are not contemplated in the same way as named subcontractors. Therefore, if the employer wishes to name a supplier, they can usually only do so by specifying the supplier in the specification in which case the contractor is bound to use that supplier for the specified goods or materials. There may be good reasons to specify a supplier, but unless it is absolutely necessary, this practice should be avoided. Once the supplier knows they are the only one in the race, the price tends to go in one direction (and that's not down). Additionally, the practice may be contrary to competition and fair trade or anti-trust legislation.

Nominated suppliers

As the name implies, nominated suppliers are also selected by the employer to supply certain goods and materials. Under the ICE conditions of contract for example (now rebranded as the Infrastructure Conditions of Contract (ICC)), there is no distinction between nominated subcontractors and suppliers and the contractor may make reasonable objection to either. The rules relating to nominated suppliers in JCT98 (i.e. clause 36) have been deleted in their entirety from the 2005 and later editions of the contract.

2.6.2 General and special attendance

Once a subcontractor has been chosen and appointed, the payments to them are directed through the main contractor, unless there are special reasons why payments should be made directly to the subcontractor by the client (for example, the insolvency of the main contractor). As we can see from the sub-section headings of this chapter, construction costs comprise:

a) labour costs
b) material costs
c) plant and equipment costs
d) general site overheads (preliminaries)
e) head office overheads

To which an element is added for:

f) profit.

When a main contractor subcontracts part of the works to another firm, they effectively pass over the costs of (a), (b) and most of (c) to the subcontractor. Therefore the costs to the main contractor of this section of work will be:

a) subcontractors costs
b) any specialised equipment or service required by the subcontractor (called special attendance)
c) proportion of general site preliminaries used by the subcontractor (called general attendance)
d) profit that the main contractor would have made on the subcontracted work.

General attendance

The CIOB Code of Estimating Practice quotes SMM7 and lists items of general attendance as being:

- use of temporary roads, paving and paths
- use of standing scaffolding
- use of standing power-operated hoisting plant
- use of mess rooms, sanitary accommodation and welfare facilities
- provision of temporary lighting and water supplies
- providing space for subcontractor's own office accommodation and the storage of their plant and materials
- clearing away rubbish.

All of which are provided on the project anyway for the use of the main contractor.

Special attendance

Other specific requirements, which may not be provided by the main contractor for their own use, are considered as special attendances, and include:

- special scaffolding in addition to the main contractor's standing scaffolding
- temporary access road and hardstandings which are required for specialist works of the subcontractor

- unloading, distribution, hoisting and placing in position, any specialist materials or equipment of the subcontractor
- special covered storage and accommodation, including lighting and power
- power supplies
- any other attendance not included in general attendance or in the above list.

All of these would not be provided on the project, if it were not for the needs of the subcontractor.

2.6.3 Prime cost sums and provisional sums

Prime cost sums (PC sums) are a procedure to include the cost of a nominated subcontractor or a nominated supplier into the main contractor's tender figure and contract sum. They are rarely used now in the UK as modern standard forms of contract do not recognise the concept of nomination following the judgements in various legal cases. The subcontractor's costs would be covered by the PC sum and the main contractor was entitled to add a percentage profit to this sum and also allow for general or special attendances on the subcontractor as described above. When the nominated subcontractor's final account was received, this replaced the PC sum in the main contractor's final account.

Provisional sums were originally sums included in the BOQ for work which had not yet been fully designed but an allowance is required in the contract. Contingencies are included in project costs as a provisional sum. Provisional sums have now taken over the role of the PC sum, in that organisations which would have been covered by a PC sum are now covered by a provisional sum, such as statutory undertakings – those organisations who are the only ones allowed to do certain work, such as connecting to the mains electricity, mains gas etc.

As the main contractor is also responsible for the programming of the works, having an amount of work in the project which is described as provisional means that they cannot fully programme all the works if they don't know the total extent of it. SMM7 therefore separates provisional sums into two categories. *Defined* provisional sums mean that the contractor has been deemed to include the scope of the work as part of their programme and therefore cannot claim an extension of time or loss and expense as a result of the architect or contract administrator firming up the actual scope, even if it increases the cost in the final account over and above the provisional sum. *Undefined* provisional sums on the other hand mean that the contractor is deemed to have not included the extent of the works in their programme and therefore may be able to claim for additional time or cost. Not surprisingly, many contracts state that notwithstanding these definitions, all provisional sums are to be regarded as defined, thus transferring even more risk to the contractor.

Dayworks will normally be included as a provisional sum and is intended to be used to value work where no other method is appropriate, i.e. as a last resort. Dayworks rates will be included for labour, plant and materials with the contractor inserting their all-in rates (i.e. including all statutory on-costs) and then add a percentage addition, which is intended to cover the disruption of taking labour from their planned activities to work on the daywork instruction at short notice.

2.7 Preliminaries and general site overheads

The RICS New Rules of Measurement (NRM) splits preliminaries into three parts:

Part 1 – Project information and employer's requirements
Part 2 – Main contractor's cost items
Part 3 – Work package contractor's preliminaries.

2.7.1 Project information and employer's requirements

The project information sections offer the tendering contractors a general description of the project, including the name, nature, location and duration of works; other general information is provided which is intended to give the tenderers a flavour of the type of work they will be taking on and which may have an effect on the tender price or proposed method of work. For example:

a) The drawings used to prepare the bills of quantities (if BOQs are provided) together with any other documents used for pre-construction information. Under health and safety laws in the UK, the client is obliged to provide the tendering contractors with all information used to develop the design.
b) The nature of the site and any existing buildings currently on the site. If the site has been developed previously (and most have been), there will most likely be existing services under the ground and the ground itself may be contaminated from previous occupation. In all these cases, it is important for the tendering contractor to know as much as possible, so they can make allowance in their tender and also their method statement and health and safety plan.
c) Any physical limitations on the site such as restricted access or site parking for the workforce or materials deliveries.
d) Any other works being carried out on the site at the same time.

The employer's requirements section basically describes to the tendering contractors how the project is intended to be managed by the employer or its consultants. Although the contractor is reasonably free to choose both the method and sequence of construction, the management and supervision of the construction will be the responsibility of specialist consultants appointed by the employer (client). Therefore, procedures must be put in place for:

a) how progress of the works will be managed and controlled
b) how costs will be controlled (i.e. costs to the client – payments to the contractor; the contractor's own costs are their own responsibility)
c) maintaining accuracy in setting out and surveying
d) how standards of workmanship (quality) will be maintained and controlled
e) approval of materials including any requirement to produce samples or mock-ups
f) notifying and obtaining appropriate certificates from statutory bodies, such as water, electric and gas authorities
g) rejection of defective work by the contractor or any subcontractors
h) requirements during the defects liability period after the works have been substantially completed

i) protection of the works, both during construction and after sections have been completed but not yet handed over to the client.

As mentioned above, the contractor is normally responsible for the method, sequencing and timing of the works during the contract period (i.e. between the date for commencement and the date for completion stated in the contract). However, there may be circumstances where limitations are put on the contractor's choice, for example, deliveries may be restricted to early morning in a busy city centre site, or a tower crane is not allowed to encroach over neighbouring land. The working hours may be restricted to prevent noise or protect neighbours' privacy or for cultural and religious reasons at certain times of the year. Any such restrictions should be described in this section to allow the contractor to make appropriate allowance in their tender price. Furthermore, if temporary works are required (such as temporary roads or protective works) and the contractor is not at liberty to design or choose the location themselves, this must also be stated.

On many larger projects, the employer and their consultants will need office accommodation on the site, suitably furnished and ready for their occupancy to supervise and manage the works. If this accommodation for the 'client-side' is required to be provided by the contractor, it must be described and included in this section. The employer must specify the number of desks required, together with IT equipment, computers, printers, telephones, fax machines etc. in accommodation which is heated/air conditioned with kitchen, toilet and welfare facilities. All such costs will be reimbursable under the contract together with the similar costs of the contractor's accommodation.

The operation and maintenance of the finished building is normally outside the contractor's scope of works. However, under health and safety laws, the contractor is obliged to pass over to the client a health and safety file, which includes all operation and maintenance manuals provided by the manufacturers for equipment installed under the contract. It is also not unusual in international contracts for the contractor to provide training for local staff in operating the building and its facilities. Details of such obligations and requirements should be included in this section of the preliminaries to allow the contractor to include the costs.

2.7.2 Main contractor's cost items

As the preliminaries section of a bill of quantities is for the pricing of items which are not 'quantity-related', i.e. they may not change in direct proportion to the quantity of work being carried out, then the main contractor's cost items which are related to time, or the method of work, would be included in this section. These costs are normally one-off payments (called fixed charges) or regular payments over the course of the contract (called time-related charges) and would include:

a) The contractor's management and staff, who are specifically allocated to this project and are not included in the all-in rate calculations contained in the unit rates for measured work.
b) The contractor's site accommodation specifically for the sole use of this project, which would include site offices, kitchens, welfare facilities, compounds, stores, site workshops etc.

c) All temporary works required to set up the site establishment, such as temporary drainage, site roads, car parks etc., including reinstatement on completion of the project, although in most cases the area of site offices will form a landscaped area in the project site.
d) All furniture and equipment (termed FFE in many contracts – fittings, furnishings and equipment) required for the temporary contractor's offices including any necessary maintenance. This is not the FFE which may be required to be provided for the contract, which will be included in the measured works of the bill of quantities.
e) Temporary utility supplies including IT and telecommunications systems including necessary maintenance and consumables.
f) Site security costs including staff, equipment, hoardings and fencing. The contractor is responsible for the site from the date of commencement of the contract until the client takes over the facility at practical/substantial completion. Therefore, site security is an important issue for health and safety and will certainly affect the insurance premiums paid by the contractor.
g) Mechanical plant and equipment which is used across the site and cannot therefore be allocated to a particular item of work in the measured work section. This particularly relates to tower cranes, mobile cranes and hoists, which will be used on the majority of sites, but will also include concrete batching plants etc. if the contractor decides to produce the concrete on site in this way.
h) Scaffolding and other temporary works which are necessary to complete the permanent works. Most construction projects use scaffolding to some degree and the erection costs (fixed charge), weekly hire costs (time-related charge) and dismantling costs (fixed charge) would be included in this section. In civil engineering projects, temporary works can be a very significant cost relative to the cost of the permanent works. For example, the caissons and cofferdams necessary to build a bridge over an estuary or river require some very specialised engineering design and construction in their own right, although strictly speaking they are temporary works required in order to build the permanent bridge.
i) Testing and commissioning costs prior to handover at practical/substantial completion, depending on what is required in the contractor's scope of works.
j) Costs to the contractor of all insurance, bonds, guarantees and warranties required under the contract.

These contractor's general cost items can be a significant proportion of the overall costs of a contract, ranging from anything between 8 per cent to 20 or 25 per cent of the contract sum, depending on the complexity of the project.

2.7.3 Work package contractor's preliminaries

The RICS New Rules of Measurement recognises the modern trend in construction procurement to enable an early involvement of the contractor and also to split the overall scope of work into separate work packages, as for example under management contracting or construction management procurement routes. Each package contractor will clearly have their own site-based costs as well as the site-based costs of the management contractor or construction manager. As the works package contractor(s) are contractors themselves, with similar responsibilities and obligations (and therefore costs) to the main contractor, many of the items in section 2.7.2 above will still be appropriate to them. It is also important that the obligations taken on by the main contractor are reflected in any work package/subcontract agreement, in what are known as 'back-to-back' agreements. Therefore, the work package preliminaries would include the following sections:

1 project particulars, stating the nature, location and details of the project
2 drawings and other documents provided to the work package contractor
3 the nature of the site and any existing buildings
4 brief description of the works
5 the conditions of contract for the work package, including content and use of documents
6 how the works will be managed
7 procedure for quality control
8 procedures to ensure security, safety and protection
9 any limitations on method, sequence and timing of the works
10 provision of site accommodation, services and facilities
11 temporary works required on the project
12 requirements concerning the operation and maintenance of the finished building.

Because, by definition, there will be multiple work packages on a particular project, the costs of the above items may be multiplied by the number of work package contractors engaged on the project, which is likely to increase the overall cost to the client. However, this is a very simplistic way of looking at it, since few of the work packages will be working concurrently and the cost saving created by a reduction in overall duration of the project usually offsets the small duplicated costs of work package preliminaries.

2.8 Head office overheads

Head office overheads include those costs of the contractor's business that cannot be charged to a particular contract, but are employed to oversee all of the projects that the company is engaged in, plus the estimating/tendering/bid management function for future projects. They are an indirect cost of the business as stated in section 2.1.1 above. Items included in head office overhead would typically include:

- head office maintenance and running costs
- mortgage or rent and rates of central or regional offices and workshops, storage facilities etc.
- wages and salaries of head office management, supervisory, estimating, surveying, administrative and accounting staff etc.
- depreciation costs on assets owned by the company
- legal fees
- interest on loans etc.

Because these costs cannot be allocated to any one project, it is common practice to distribute head office overhead costs to all contracts currently in progress, which can be done in a number of ways. The total amount of the overhead is ascertained from company accounts and usually adjusted on an annual basis, although as the industry becomes more competitive and streamlined, these costs are permanently under review and the percentage addition to the estimate will also be reviewed more often.

At the beginning of the trading period (annually or quarterly as noted above), a contractor will estimate their turnover for the year ahead and determine what percentage of turnover is required to contribute to the head office overhead cost. From this the estimator will be able to establish the percentage which must be added to the estimated direct cost of a contract to cover the head office overhead contribution. This percentage contribution is typically spread across all items in a contractor's tender submission. Despite being common estimating practice this process does fail to recognise that head office overhead is as much a function of time as it is of cost.

2.9 Profit mark-up

Classical economics states that profit is the reward for taking risks, and that the higher the risk, the higher the required profit. However, many directors of construction companies would scoff at such academic notions as contractors are being required to accept more of the project risks while at the same time having their profit margins reduced by increasingly aggressive procurement methods being used by clients and their advisers, such as reverse auction and on-line bidding, post-tender negotiated reductions euphemistically described as 'best and final offer' and stealth clauses such as design verification which can effectively convert the contract from full design to EPC/design and build by making the contractor responsible for the design coherence and buildability.

The decision of how much profit mark-up should be added to an estimate will depend to a large extent on the following factors:

a) the anticipated cash flow of this project within the overall cash flow of the company
b) how well defined the scope of works is within the tender documents
c) the degree of risk being passed to the contractor
d) whether head office overheads are being recovered by other contracts
e) minimum level of profit required by the directors
f) degree of discounts and profit on subcontractors
g) VAT, whether it is payable and the terms of payment.

Head office overheads and contractor's profit are unlikely to be spread across the entire project at the same rate. As with all commercial organisations, construction contractors rely on cash flow to pay their bills, salaries, wages etc. and unless there is an advanced payment, the contractor is likely to be in a negative cash flow for the majority of the project, as payments are received from the client up to two months after costs have been incurred on the site. Additionally, reductions such as retention or disputed items will further reduce the contractor's incoming cash flow.

It is therefore not surprising that contractors will want to build up some profits in the early part of the project to help offset the 'normal' negative cash flow. Early activity and trades, such as the mobilisation items in the preliminaries section together with excavation and ground works may consequently have a higher profit margin on their rates than the later trades such as decoration. If, however, the contractor does not consider the project to be a significantly adverse risk, the overheads and profit will be spread equally throughout the project. For this reason, the overhead and profit element is also known in the industry as the 'spread percentage'.

See Chapter 7 for further discussions on bidding strategy and converting an estimate to a tender.

Example

A construction company, engaged in commercial office developments prepares a financial budget for its contracts department at the beginning of each calendar year. The situation on 1 January 2011 was as follows:

Contract	*Contract sum (fixed price)*	*Period of work*
1	£ 2,400,000	1 Oct 10 to 1 Feb 12
2	£ 1,300,000	1 Dec 10 to 1 Jan 12
3	£ 1,200,000	1 Nov 10 to 1 Sep 11

The company is at present negotiating a further contract valued at £1,800,000 to commence on 1 June 2011 with a duration of 12 months.

Each of these contracts must include an allowance to cover budgeted head office overheads of £430,000 across all four projects and budgeted profit of £240 across all four projects.

The budgets are reviewed annually at 30 June, to incorporate the actual costs incurred during the first six months of the year. The actual costs at 30 June 2011 are as follows:

Project	*Value of work done in the half year*	*Direct costs to date*
1	£ 1,100,000	£ 1,000,000
2	£ 700,000	£ 640,000
3	£ 800,000	£ 680,000
4	On hold – Contract not awarded yet.	

The actual costs of head office overheads for the first six months of 2011 are £180,000 which should be split proportionately between the contracts.

Assuming equal monthly valuations throughout the contract (although this will be unrealistic as the valuations will be determined by the programming of the works and the S-curve), let us first look at the budget prepared at the beginning of the year:

There is an anticipated total income of £2,370,000 in the first half of the year with an anticipated full year income of £5,010,000. It should be noted here that these figures relate to the *value of work done* and NOT to monies received from the client. If the contractor performs £370,000 worth of work in January, this money (less a proportion for retention – usually 5 per cent) is unlikely to be received before March, depending on the valuation and certification procedures in the particular contract for the project.

Table 2.1 Worked example – total budgeted income

BUDGET FOR YEAR 2011

	Jan	*Feb*	*Mar*	*Apr*	*May*	*Jun*	*HALF YEAR*	*Jul*	*Aug*	*Sep*	*Oct*	*Nov*	*Dec*	*FULL YEAR*
Project 1	150,000	150,000	150,000	150,000	150,000	150,000	900,000	150,000	150,000	150,000	150,000	150,000	150,000	1,800,000
Project 2	100,000	100,000	100,000	100,000	100,000	100,000	600,000	100,000	100,000	100,000	100,000	100,000	100,000	1,200,000
Project 3	120,000	120,000	120,000	120,000	120,000	120,000	720,000	120,000	120,000					960,000
Project 4						150,000	150,000	150,000	150,000	150,000	150,000	150,000	150,000	1,050,000
TOTAL	370,000	370,000	370,000	370,000	370,000	520,000	2,370,000	520,000	520,000	400,000	400,000	400,000	400,000	5,010,000

Table 2.2 Worked example – total budgeted contribution to head office overheads

BUDGET FOR YEAR 2011

	Jan	*Feb*	*Mar*	*Apr*	*May*	*Jun*	*HALF YEAR*	*Jul*	*Aug*	*Sep*	*Oct*	*Nov*	*Dec*	*FULL YEAR*
Project 1	9,375	9,375	9,375	9,375	9,375	9,375	56,250	9,375	9,375	9,375	9,375	9,375	9,375	112,500
Project 2	6,154	6,154	6,154	6,154	6,154	6,154	36,923	6,154	6,154	6,154	6,154	6,154	6,154	73,846
Project 3	8,000	8,000	8,000	8,000	8,000	8,000	48,000	8,000	8,000					64,000
Project 4						10,000	10,000	10,000	10,000	10,000	10,000	10,000	10,000	70,000
TOTAL	23,529	23,529	23,529	23,529	23,529	33,529	151,173	33,529	33,529	25,529	25,529	25,529	25,529	320,346

Each one of these projects would be required to make a contribution to the head office overheads and also to the profits of the company.

If we look at how these figures have been calculated, the January figure of £9,375 represents 1/16th of the head office overheads allocated to project 1, as the total duration of project 1 is 16 months. Therefore the head office overheads are allocated to the projects on a time basis.

However, this is not the only way of allocating overheads. The company management may choose to allocate the overheads on a total cost basis, so that each project takes a 'fairer' share of the overheads, depending on the size of the project rather than just the duration. Allocating overhead costs in this way would have quite different results for the total costs allocated to each project and therefore the profitability of each project.

Consequently, by a simple management choice of allocating indirect costs in different ways, the profitability of each project (or 'cost centre' as the accountants may also term them) is fundamentally affected. This particular ruse is not unusual in corporate finance since indirect costs do not strictly belong anywhere, so they can be allocated almost at will, as long as the company stays within the statutory accounting procedures. Although not part of the scope of this book, the treatment of depreciation of plant and equipment is an even more slippery concept, since depreciation costs do not involve any actual payments, unlike head office overheads such as telephone bills. Depreciation is just an assessment by the company of the reduced value of assets over time and can be calculated in several different ways, each of which will have a different effect on the company's profits. However, the statutory accounting procedures issued by the tax and revenue authorities in most countries do give strict guidelines on how certain assets can be depreciated, to try to avoid the more creative accounting tricks which have occurred in the past.

2.10 Summary and tutorial questions

2.10.1 Summary

The study of costs and how they behave is a very complex area of economics and this book does not wish to go into the details of economic theory – there are plenty of texts that do that. What the book does want to do is try to explain in a very practical way how construction costs can change depending on:

a) the amount of work to be carried out
b) the time taken to do the work
c) the amount of labour and plant/equipment allocated to the item of work
d) the effectiveness of the management control processes.

Although costs can be divided into direct/indirect or fixed/variable, it is this first definition which is most useful to us in construction. Direct costs are those which directly relate to the project, and the critical point is that if the project did not take place, these costs would not be incurred. Indirect costs on the other hand are those costs which would be incurred irrespective of the actual project costs. Therefore all costs related to the site are direct and all costs related to the head office or regional offices are indirect. Generally, direct costs comprise the following:

- labour costs
- material costs
- plant and equipment costs.

With the actual costs varying depending on the productivity of the labour or item of plant/equipment and the different on-costs which must be allowed for, the employment of people involves considerably more costs to the employer than just wages. So all these extra costs are accounted for in the all-in rate. The same applies to items of plant and equipment owned by the company.

The cost of materials will vary depending on the commercial arrangements negotiated with suppliers (i.e. discounts etc.) as well as the efficiency with which the materials are delivered, stored and incorporated into the works. Poor stock control and high wastage factors will all increase the average costs of materials, thus increasing project costs and possibly reducing the profit margins of the contractor.

The last major item of direct site-based costs is that of subcontractors. These can either be labour-only subcontractors, in which case the main contractor is paying a profit margin on top of the all-in rate (which may be acceptable as subcontractors are not employed and therefore have no employment rights), or the subcontractor may be required to supply labour and materials for their scope of works. In both cases, the main contractor will be required to provide site facilities, which are termed ‘attendance’.

The cost of preliminaries do not fall neatly into the classification of direct or indirect costs. Strictly speaking, by the above definition they are direct costs as they will not be incurred if the project does not take place. However, another definition of direct costs is that they are the costs *of the actual work* taking place. Using this definition, they are indirect costs. However, as they are site-based costs, the first definition is usually adopted by most construction companies.

Formal indirect costs relate to the costs of the organisation at head office or regional office, including the various central departments of finance, HR/personnel, estimating, construction and contract management etc.

2.10.2 Tutorial questions

1. Give a list of examples of 'direct costs' and 'indirect costs' on a construction project.
2. How can a contractor ensure that labour is as productive as possible?
3. Look at the elements which make up a labour all-in rate. Which of these elements is under the control of the contractor and which are not?
4. Materials costs will be the same for all contractors. Comment on this statement.
5. How can a project design reduce the amount of material wastage in a project?
6. Compare the advantages and disadvantages of owning an item of plant against hiring from a specialist plant hire company.
7. Compare the advantages and disadvantages of the main contractor carrying out construction work themselves against employing a subcontractor.
8. What is the difference between general attendance and special attendance?
9. What is the difference between a defined and undefined provisional sum?

3 Early contractor involvement in tendering procedures

3.1 The move to early contractor involvement

Early contractor involvement (ECI) is about engaging the contractor during the early stages of a project in order to contribute to the design development and hopefully to promote a better understanding of the risks and responsibilities by all of the parties to a project. It is not a tendering procedure or procurement route itself, but is an all-encompassing term for those procurement routes which require the contractor to be appointed at an early stage in the project when their technical and management skills can make a significant and beneficial contribution to the design development.

In contrast to the traditional procurement route where the contractor is appointed after the design has been completed and therefore only having an involvement in the project during the construction stage, ECI involves the contractor working with the client and design consultants in the initial stages of the project to develop the design and a detailed project plan. In order to get an indication of the construction costs and, therefore, how much the contractor will be paid for the works, the parties will also develop an outline price for the project, which is effectively an approximate estimate as defined in the previous chapter. Although similar to a design and build arrangement, ECI has the added benefit that the outline price is not fully agreed until all the risks can be properly assessed and allocated to the parties. The outline price will normally be converted to a lump sum fixed price before construction starts in earnest.

3.1.1 Advantages of ECI

The tendering process for ECI projects is likely to be less intensive and less costly for the client, for which they will no doubt be pleased. It is aimed at selecting the best team to deliver a project and consequently does not require the tendering contractors to prepare detailed lump sum quotations for the actual construction stage of the works.

Other advantages include:

- a shortened delivery time, through concurrent working, as construction may commence while the design is developing
- a team approach, as the builder is part of the initial project team and not seen as an adversary, which can be the case in traditional procurement
- the contractor's skills and experience is harnessed early in the project
- increased opportunity for innovation and buildability
- quicker decision making

- better integration of construction methods, by using the contractor's technical and programming/scheduling skills
- earlier procurement of materials, which is particularly important for items which require long lead times
- fewer variations during construction, as the design is more likely to be robust, practical and efficient
- fewer defects during the life cycle of the building as the contractor can identify and eliminate weaknesses in the design which can lead to weaknesses in the structure and fabric of the works.

However, it is not all sunshine and plain sailing. The difficulties which may be encountered with ECI projects may include:

- a heavy involvement of senior staff from all participants in the early stages of the project for longer periods
- additional costs through assessing design alternatives and value engineering workshops throughout the design development stage
- the possible need for independent cost estimators to prevent inflated prices resulting from the non-competitive building up of the construction costs.

The following article highlights these and other issues of ECI especially during the recessionary period for the construction industry.

How to be good when times are bad: early contractor involvement

By David Mosey

Early contractor involvement, both on single projects and through frameworks, has increased significantly during the 10 years of relative prosperity. The question is: can the systems governing this form of procurement stand up to the demands placed upon them by recession?

Commentators on partnering, which is closely linked to early contractor involvement, have suggested that it can only survive in the right economic climate – one example being the initiative between main contractors and subcontractors in France in the mid-eighties that collapsed in the 1988 recession. However, the French did not have the sustained government and industry support for early contractor involvement that has been generated in the UK and illustrated through a wide variety of case studies.

For example, the 2005 National Audit Office report, Improving Public Services Through Better Construction, describes the experience of the University of Cambridge. It used to use single-stage contractor appointments in 1998, and it achieved costs 2 per cent above budget and eight-week delays. When it changed to an early contractor involvement model in 2002, its costs were 3 per cent below budget and there were no delays.

The cynics will say that early contractor involvement is too cosy, that it takes the edge off competitive pricing and that clients cannot afford to risk such an approach in strained economic times. The answer to the cynics is that if procurement and contractual systems are set up properly, then early contractor involvement gives far better cost and programme control. It provides a way to get the contractor's ideas about the scope of the project and its supply chain, and there is still plenty of thinking time before the contractor is authorised to start work on site.

And what about the inevitable tensions that build up during a project? Where does early contractor involvement stand then? I have recently been involved in settling two prospective disputes in Bahrain and Dubai, both of which arose on projects procured through early contractor involvement. In each case, the detailed cost and time data obtained through an open-book build-up of the price and programme, as well as the clear points of interface between consultants and contractors, allowed the parties to step away from adversarial positions and ultimately to settle their differences without going to court. Properly structured early contractor involvement should offer the same opportunities in the UK.

As to the structures available, PPC2000 and PPC International are specifically set up for early contractor involvement. The September 2008 Arup report for the Office of Government Commerce described PPC2000 as 'a complete procurement and delivery system that is distinct from other forms of contract available'. Meanwhile, NEC3 has been adapted for early contractor involvement by clients such as the Highways Agency, and JCT has published its own Preconstruction Services Agreement.

So what are the features of properly structured early contractor involvement? They should include:

- Consultant and contractor selection on the basis of their ability to deliver value, including early agreement of costs where possible and avoidance of percentage profits/fees that encourage cost increases.
- Main contractor appointment on a conditional basis until designs and prices are agreed. This should be enough for it to contribute to designs and to build up second-tier prices through open-book subcontract tenders.
- Sufficient time for value engineering and risk management exercises that allow the contractor and its prospective subcontractors to help iron out inappropriate costs or unaffordable/unbuildable designs.
- Binding programmes for deliverables for all consultants, contractors and subcontractors throughout the preconstruction period.

And what should early contractor involvement avoid? Most importantly, the opportunities for brinkmanship and last-minute negotiation that arise if agreed preconstruction phase processes are not rigorously followed. It should also avoid separating the preconstruction appointment and the construction phase appointment so as to calm the fear of contractors that their best ideas will be grabbed by the client and put out to tender for someone else to build. Third, the agreed structure should avoid the risk of misunderstandings by linking consultant and contractor appointments around single integrated programmes.

It should be remembered that the recommendation of early contractor involvement in the 1994 Latham Report was not born in a period of economic

> prosperity, but during the last recession. However, the merits of the system will be tested during this recession and it will only stand up if structured to deliver the savings and efficiencies cash-strapped clients demand.
>
> ---
>
> Article in *Building* magazine dated 13 March 2009

So, involvement of the contractor early in the project, when the design is being developed, has many advantages and some disadvantages. As far as the content of this book is concerned, from an estimating point of view, the initial cost estimates will be a mixture of 'top-down' techniques taken from historical cost data and 'bottom-up' techniques using primary resource costs. This is because the cost estimates will be mainly generated by the contractor, who has historically used project scheduling techniques and resource costs to build up their estimates and tenders.

In terms of how the contractor is appointed on ECI contracts, this would normally be through a two-stage process, possibly requiring two separate contracts (hence the standard description of 'two-stage' tendering). The first stage is effectively a consultancy contract, preferably through an NEC Professional Services Contract (PSC) or through a bespoke Pre-Construction Services Agreement (PCSA), either of which would be used to appoint a suitable contractor who is selected through a tender process based on quality and price, to arrive at the most economically advantageous tender (MEAT). Individual prices for labour, plant, overheads and profit would all be required for the evaluation of this initial part of the tender process. These prices, along with material costs, would also be incorporated into the second stage after the design is completed, to calculate the fixed price construction costs, which are often set out as a 'target price' with penalties or bonuses (painshare and gainshare). Under the first stage PSC contract, the contractor is obliged to provide input into the design process, instigate procurement of work package contractors and address the various risks inherent in the contract in order to work towards either a lump sum or a target price for the second stage, the construction phase. If a lump sum or a target price is reached which is acceptable to both the contractor and the employer, then the contractor is appointed under a target contract, to construct the scheme. Should an acceptable target not be agreed, or sufficient funding is not available at the time, the employer should ensure that there is a clause which enables the termination of the overall contract with no financial effect to the employer.

3.1.2 Disadvantages of ECI

It is generally accepted in the industry that the advantages of ECI far outweigh the disadvantages, although contractors have expressed concern that ECI contracting ties up senior staff for much longer than under more traditional forms of procurement – staff who should be on site earning turnover, thereby contributing to recovery of head office overheads. This additional involvement occurs particularly in the early stages of the design phase when (under usual ECI forms) contractors are often reimbursed at hourly rates over an extended period. But this work has relatively low profit margins and contractors traditionally make their main profit by taking on the risks in construction. It is probably too early to tell whether in the long run this issue will outweigh the benefits of enhanced employer relationships, which ECI is designed to deliver and hopefully these benefits will make up for the intensive up-front managerial staff involvement.

A further disadvantage is that ECI limits the opportunity for competition on price, since fixing a competitive lump sum price so early on in the project is clearly not possible. This risk can be limited by deferring the formal appointment until the contractor can guarantee a maximum price. The early selection of a contractor therefore carries one very significant caveat – that the process must be even more transparent than when the contractor is selected at a later stage in competition with others. Competition and anti-trust legislation in many countries usually requires that all prospective contractors have equal opportunities in tendering. When the selection is done on price alone, the process is inherently transparent; when selected on other criteria, such as value, the decision process has to be made as transparent and unbiased as possible.

Therefore, early involvement of the contractor means that the project is no longer in competition and, as a result, despite the employer and their consultants trying to control the procurement process, they may find that anticipated project costs rise throughout the pre-construction period. In general, the employer's commercial bargaining position will be progressively weakened the longer the contractor is involved without a formal building contract being agreed and signed. Once a contractor becomes integrated as part of the project team and has carried out extensive pre-construction services, the employer may be reluctant to dismiss them in the event of failure to perform or a refusal to negotiate sensibly on price and programme. Whilst the pre-construction agreement may be drafted to allow the employer to terminate the contractor's employment, it is important that employers realise that if they allow the contractor to go too far into the design and/or construction programme, it may be too late to stop or turn around. It is not uncommon, as part of the pre-construction services, for a contractor to carry out some early enabling or investigative works on site by a mini-contractor letter of intent. However, it is advisable not to go beyond this stage. If the contractor is allowed to carry out more detailed or permanent works on site before agreeing the terms and conditions of the substantive building contract, then the employer's negotiating position will be considerably weakened.

Let us now take a look at the main modern tendering procedures with specific reference to how estimates are generated.

3.2 Fixed price (lump sum) tendering

This method of procuring building projects is usually referred to within the industry as 'the traditional method' since it accounts for over 60 per cent of construction work in the UK, and over 90 per cent in many other parts of the world, such as the Middle East. As well as the total separation of design and construction, the traditional procurement system has other characteristics:

1 Project delivery is a sequential process, i.e. there is no overlap of the functions, so construction cannot start until the critical design has been completed, as the contractors are appointed on the assumption that there is sufficient design on which to base a competitive price and that any balance of design will be delivered in sufficient time to enable the contractor to proceed unhindered.

2 The responsibility for managing the project is divided between the client's consultants and the contractor.
3 The contractor is paid for the work completed on the basis of either a proportion of the bills of quantities, or on completion of activity milestones.

At the beginning of the project, the client appoints independent professional consultants, who fully design the project, with the exception of any performance specified work or contractor's design portions and prepare tender documents upon which lump sum competitive bids are obtained from interested construction firms. The successful tenderer enters into a direct contract with the client and carries out the work under the supervision of, usually, the original design consultants.

3.2.1 Preparing and obtaining tenders

Tender documentation on traditionally procured projects normally consists of:

- tender drawings
- specification(s)
- pricing document (usually a bill of quantities, but may be an activity schedule).

The bill of quantities is prepared by the consultant quantity surveyor from the designers' drawings in order to provide each tenderer with a common base on which to price their bid. For the traditional system to operate successfully, and to minimise the financial risk to the client, it is imperative that the design is sufficiently developed before the bills of quantities are prepared and tenders invited. If this is not done, the contractor's estimator may be prevented from calculating an accurate price for the works and excessive variations and disruptions of works are likely to occur during the construction stage.

The selection of the contractor for the works is generally made by selective tendering based upon a list of tried and tested contractors whose past performance, financial stability and resources have already been established and in most cases, regularly monitored.

Once tenders are received, the selection of the best bid is fairly straightforward if it is judged on price alone, having been based on documentation which is common to all tenderers and which, theoretically at least, accurately and comprehensively reflects the client's actual requirements. However, care must be taken at this point as the bids will often only be a single lump sum. For example, the form of tender may only state:

> We, XYZ Construction Company Ltd. Offer to carry out the Works known as XXXXX in accordance with the tender documents, for the sum of £876,543.00.

This may be the lowest bid, but how do we know it is a serious bid that has been properly estimated? It is therefore important to conduct a tender analysis where the lowest two or three tenders are called in for technical and arithmetic checking. The National Joint Consultative Committee (NJCC) for the construction industry, which was made up of representatives from all sides of the industry published a series of documents on the Code of Procedure for Tendering and although the NJCC no longer exists as a consultative committee, the Codes are still an extremely useful guide for how tenders should be managed, evaluated and chosen. The Code of Procedure for Selective Tendering recommends that the three lowest tenders are checked before finally selecting the successful contractor.

When using the traditional system of procurement, the successful contractor should be given an adequate period of time to plan the project thoroughly and organise the required resources. Undue haste in making a physical start on site may result in managerial and technical errors being made by both the design team and the contractor, which could lead to a lengthening rather than a reduction of the overall construction period.

A major disadvantage of the traditional route is that a very high proportion of the estimated cost of the project has been committed before work commences on site, although actual expenditure is comparatively small. This is because costs are generated by design decisions, but they are incurred when the work is carried out. However, it is during the construction phase that the majority of difficulties will emerge with the quality of the performance during this period having already been largely determined by the quality of design decisions. It is at this stage that the price for an incomplete design, inaccurate bill of quantities, poorly prepared tender documentation and lack of 'buildability', etc. is paid. An example is the Harrogate Conference Centre in North Yorkshire, UK. In this project, the foundations were designed and the contractor appointed on a traditional contract. However, because the site ground conditions were not checked thoroughly at the design stage, a geological fault line was found to traverse across the site, requiring a complete redesign of the foundations to 'overarch' this fault line, with the consequent increase in project duration (and loss and expense costs to the contractor). Therefore, by not completing a full ground survey and design, the actual costs were significantly more than originally planned and the construction period lengthened while the foundations were redesigned and constructed.

Payment to the contractor for work that has been satisfactorily completed is made by means of interim certificates, generally monthly, but as mentioned above, may also be on the basis of activities completed, to the value of work done, issued by the architect or contract administrator on the recommendation of the quantity surveyor. The priced bill of quantities submitted by the contractor at tender stage forms the basis of these interim valuations and also ensures that any variations can be valued by reference to pre-agreed rates for appropriate items of work. An agreed percentage is usually retained until practical or substantial completion of the works, when a portion (usually half) is paid to the contractor. The remaining half of the retention fund is paid at final completion after the defects correction period.

The major selling point of the traditional system and one which is made by countless consultants across the world is that, by using bills of quantities as part of the tender documentation, the cost of tendering is reduced, the quantitative risks encountered in tendering are removed, competition is ensured, post-contract changes can be implemented at a fair and reasonable cost and clients can be confident that they know their financial commitment. All this is true, provided that the design has been fully developed and accurately billed before obtaining tenders. If, however, these criteria have not been strictly met, excessive variations, disruption of the works and a consequent increase in the tendered cost will occur. Also, there is often a considerable amount of work covered by provisional sums which will necessarily vary when the actual extent and cost of the work is known, so the final account may be quite different from the original tender lump sum or contract lump sum.

3.3 Design-build and management contracts

In design-build arrangements, the contractor takes responsibility for both the design and construction of the building, even though they may bring in external companies to design the work, in the same way as they bring in external companies for the air conditioning installation. The designer is as much a subcontractor as the MEP firm. The client merely provides the initial 'employer's requirements' (ERs) and the design-build contractor is responsible for fulfilling these requirements. Although the client places the design of the works in the hands of the contractor, a considered development and presentation of the ERs will enable them to retain a strong control on what is being designed and will also enable a great deal of risk to be passed to the contractor. This form of procurement is becoming the most popular in use today, particularly with private developers who are the most keen to pass on as much of the risk as possible.

3.3.1 Engineering, procurement and construction (EPC) contracts

This type of contract is more common in process engineering projects such as those found in the oil and gas industry where the project is more accurately defined by functional performance criteria than by aesthetic design criteria, which is more common in the building and civil engineering industry.

At the inception of the project, the front-end engineering and design (FEED) is the stage which generates the basic or conceptual design (i.e. up to Stage C in the RIBA Plan of Work – see Table 4.1 in Chapter 4). This FEED stage is usually executed on a schedule of rates basis, although certain definable components can also be let on a lump sum basis. The FEED stage usually includes:

- basic engineering and design
- project schedule and cost estimates for project control
- procurement of certain long-lead items of equipment.

The FEED stage results in a basic engineering package or packages, which ideally are sufficiently progressed to enable competitive bids to be obtained on a lump sum basis.

The FEED may be produced in-house by the client organisation, if they are sufficiently experienced, or by specialist engineering consultancy firms. However, nowadays many EPC contractors have the in-house capability to carry the work out themselves, which makes the contract a more pure design-build arrangement. It is also important to note that it is not really possible or desirable to define what constitutes basic engineering or FEED in general terms – it will vary from project to project as well as between sectors. Across different projects and industries the FEED can be developed to a greater or lesser extent, sometimes getting close to or constituting an element at least of detailed design.

3.3.2 Engineering, procurement and construction management (EPCM) contracts

This could be described as being primarily a professional services contract, although the procurement structure is more similar to a typical construction management approach but with the vital difference that the detailed engineering and design function is normally carried out by the EPCM contractor.

The EPCM contractor is responsible for:

a) development of the design from the schematic design included in the FEED documents
b) procurement of necessary materials and equipment
c) management and administration of the separate construction contracts.

An important difference between the EPCM and EPC contracts is that in the EPCM model, the contractor is providing professional services (including design) and is not a party to any of the construction contracts. The EPCM contractor acts as the employer's agent/client representative by arranging the contractual relationships between the client and trade contractors. Each trade contract is formed between the client and the trade contractor. Ideally, in such an arrangement, as with typical construction management, the client would be well advised to have a large and experienced in-house team to assist the EPCM contractor with the management and administration of these contracts. Whilst this is strictly speaking a very large part of the important role for which the EPCM contractor is paid, given the division of responsibilities between various parties in an EPCM arrangement, it is vital that the client keeps a careful eye on performance of each. The EPCM contractor will not usually take responsibility for delivering the completed project by an overall completion date (thus rarely are there liquidated damages provisions in EPCM contracts for delay to the project as a whole), nor will it take responsibility for care of the works or for the ultimate cost to the client of the project. However, incentives can and are often built in to the EPCM contract in this regard (see below).

Consequently, the principal functions of the EPCM contractor relate to:

a) the performance of the design work
b) the preparation of the budget cost estimate
c) the preparation of the estimated duration of the work
d) managing the procurement and administration of the trade contracts
e) coordination of the design and construction between the trade contractors

3.4 Call-off contracts – partnering and framework agreements

A partnering or framework agreement is simply a time limited (or 'term') contract between a client and a contractor that governs and overarches individual project agreements awarded under it during the term of the agreement. It may be used where a client body, knowing it has a number of construction projects to implement over a time period, decides to carry out a single tender process for both the consultants and contractors that it will need for the projects. Alternatively, it may be used by a client who wishes to develop a stronger relationship with a particular single contractor or consultant across a number of its projects. Whilst the framework agreement is generally between the employer and contractor, there is a positive encouragement for the contractor to enter similar framework agreements with its own suppliers and subcontractors, thus including the entire supply chain within the overall framework. As each individual project is required, the underlying agreements are 'called off', with the terms and conditions having already been agreed.

For the client, the framework agreement has the obvious benefits of reducing costs and creating certainty in the costs of work during the overall period of the agreement. If it can carry out one tender exercise for ten projects, rather than tendering each project separately, there will naturally be both a cost and time saving. The client also has the benefit of choosing their preferred contractor from the framework based on availability of experienced and trusted teams.

There is also the possibility of economies of scale. The call-off framework agreement will often require the supplier (contractor or consultant) to include in the overall tender a price mechanism for the individual project contracts that may be made under the framework agreement. Where the contractor or consultant is likely to obtain a steady stream of work over a period of years, this learning curve is expected to be reflected in the price. The framework agreement will normally require the contractor to improve their performance over time as they become more familiar with working with the client or on a particular type of project. This performance will be monitored and judged by benchmarks and key performance indicators (KPIs), or alternatively, a percentage reduction in preliminary costs over a number of projects may be required or negotiated. As in any relationship, communications are expected to improve over time as each party becomes more familiar with how the other works and what is important to them. Whilst this can be obtained through individual contracts over a period, the improvement should be more structured and measurable within a more formal framework agreement.

3.5 Reimbursable (cost-plus) contracts

Cost reimbursable (or cost-plus) contracts are not uncommon on construction projects, and are used when the conventional lump sum, fixed price contracts are not considered flexible enough to deal with the requirements of a particular project, mainly an immediate start on site. However, there are two major cost weaknesses of this type of contract; first, the lack of knowledge of their overall financial commitment by the client and second, the lack of incentive for the contractor to control their costs, since they are paid everything they spend on the project. The lack of knowledge of overall financial commitment is clearly related to the lack of definition of the scope of work at the tender stage, but the client uses a cost-plus arrangement because they want the work done sharpish.

However, cost reimbursable contracts also have many advantages for both the client and contractor.

a) Design and construction can progress simultaneously, leading to an early completion and usage of the facility by the client.
b) In the case of a revenue-earning facility, income from rentals or sales can be generated earlier than with other forms.

In order to achieve these benefits, an effective and practical cost and schedule control system must be established for the project in order for delivery to be on time and within budget. Such a system should also be able to provide the information required by the project team to compare the actual progress with the planned progress and use the techniques of earned value management (EVM) or variance analysis.

Consequently, the project team can verify whether work is in line with or deviates from the original plan. This also highlights contractor performance and any labour and plant productivity, indications that are essential for keeping control of the project and, if necessary, for identifying corrective measures. In order to control cost-plus contracts effectively, this 'benchmarking' is an essential requirement.

Therefore, in cost reimbursable contracts the contractor is paid their actual costs including preliminaries together with a fee to cover their overheads and profit which may be either a percentage fee or fixed fee. As stated, this payment mechanism is appropriate where an early start is required but the project lacks sufficient definition to allow a fixed price or firm rates to be established.

Cost reimbursable contracts create shared risks between the client and contractor and therefore have a major effect on the relationship between the parties. The main part of the financial risk clearly rests with the client as they have to fund all actual costs of the project and have no accurate indication of the final out-turn costs. This means that the contractor may have little incentive to work efficiently and economically, unless there is a benchmarking procedure in place, or they are interested in further work from this client. It is therefore in the interests of the client to ensure that the contractor is encouraged to cooperate in forecasting the final out-turn costs, so that joint action may be taken at the appropriate time to prevent any cost overrun.

There are two main ways that this can be achieved: first, to create a legal relationship which requires the contractor to notify the client when they have reason to believe there will be a cost overrun. This is the approach adopted in the USA, where doctrines of good faith and fair dealing have developed beyond those in the UK and elsewhere through very highly developed anti-trust legislation. The second approach is to share the risk of cost between the client and contractor by using target cost contracts or maximum guaranteed price. The former is the most common form of cost reimbursement contract in UK construction.

3.6 Summary and tutorial questions

3.6.1 Summary

There are a considerable number of advantages in engaging the contractor during the early stages of a construction project, and although there are also some disadvantages, the main reason why it is not done more often boils down to the 'traditional' way of doing things in the industry and a somewhat specious notion that lump sum tendering based on detailed design somehow gives the client greater control of project costs. As described in Chapter 5, the estimating techniques used at the early stages of a project are becoming increasingly more sophisticated and also increasingly more accurate, so there is little difficulty in using these costs as a basis for the contract price with the contractor. This is in fact what happens with many process engineering plants, which are awarded under EPC terms following a schematic front-end design.

Additionally, many projects are awarded on management contracts, i.e. where the 'main contractor' subcontracts all of the work to specialist firms. In this case, the management contractor (or EPCM firm) would be involved at an early stage and contribute to the cost management of the whole project.

In terms of estimating techniques, if a contractor is involved early – before the design is completed – the approximate estimating techniques take on a greater contractual significance than if the contractor was appointed more traditionally, where the resource-based estimates would be more appropriate. Either case is perfectly valid, as long as the parties understand how the project risks are shared out.

3.6.2 Tutorial questions

1 What do you understand by the term 'early contractor involvement'?
2 List the main benefits to a client of early contractor involvement.
3 Outline the difficulties of establishing a fixed price for a contract before the design has been sufficiently completed.
4 What are the essential differences between design-build and management contracts in terms of construction cost estimating?
5 How are costs established in framework agreements?
6 Discuss the various financial risks to both the client and contractor of a cost reimbursable arrangement.
7 What is the purpose of a standard method of measurement as far as estimating is concerned?

3.7 Case study – harbour and marina works

3.7.1 Introduction and background

Why choose a 'cost-plus' arrangement at the start of the project?

'Cost-plus' is not normally the first choice of contractual arrangement, but is useful when, for reasons of urgency and in the absence of a sufficiently detailed design, it is necessary to start to the works as soon as possible.

Back in 2007, the design was indeed in its early stages and the objective was to complete the works on a fast-track basis so that the project would be completed in September 2009. Given this situation, a cost-plus contract was therefore the only possible way to start the works.

Based upon the concept design and the information known at that time on the project, a constraint that the joint venture main contractor (referred to as JV-MC) could not or should not have ignored, the contract was signed in October 2007. Its price was based upon a preliminary estimate of the estimated maximum cost (US$485 million), to be subsequently refined into the final estimated maximum cost upon completion of the detailed design.

The commitment to perform the amount of works of such magnitude and in the period of time allowed implied that the JV-MC had to mobilise exceptional means and resources; it also implied that the JV-MC had to prepare an elaborated but realistic time schedule to meet the stipulated date for the execution of the contract. Besides, the JV-MC had to revise the estimated maximum cost when the design would be substantially finished; for doing so, the bill of quantities, with the unit prices agreed by the parties in the preliminary budget, had to be updated on the basis of the adjusted quantities.

Why was it necessary to change the form of contract?

As the project developed, it appeared that the JV-MC had either under-estimated the underlying constraints of the commitment they had taken up or were not willing to deploy the necessary resources to meet the deadlines opposed. In spite of numerous reminders by the client and engineer, the JV-MC proved to be more efficient in justifying its inability to control the overall project costs and delays, rather than to take the necessary steps to abide by the fast-track nature and constraints of the project. It thus appeared necessary to change the contract set-up.

In the meantime, when it became clear that the completion date of September 2009 would not be reached, the client agreed to compromise on this objective provided that the JV-MC would make a reasonable proposal for a new price as maximum guaranteed price (MGP).

Therefore, as from early 2008, discussions started between the client, with the assistance of the engineer, and the JV-MC to obtain a quotation for an MGP/lump sum price. After almost one year of discussions, however, it appeared that under such form of contract no common ground of agreement could be reached, both in terms of prices and conditions related to such prices.

Therefore, the only way out of this dilemma, and taking into account the need to have a better control of the JV-MC's expenses, was to discuss a new form of contract where all costs/prices would be stated in full transparency. The appropriate form of contract appeared to be 're-measurable and unit prices', i.e. a form of contract with (i) unit prices for the completion of concrete works and some other works to be performed by the JV-MC itself (e.g. some finishing trades) and (ii) lump sum or unit prices for the works packages to be subcontracted, with a margin for management of such subcontracts (as is usual for a general contractor).

The negotiations for the definition of such new form of contract began during the second term of 2009. The first step of this process was a Memorandum of Understanding (MOU) spelling out the main principles that would govern the Addendum to the Contract and a time frame for such agreement. The discussions over the MOU began at the end of June and it was formally signed on 3 August 2009, with a period of two months agreed upon for the finalisation of the Addendum to the Contract.

3.7.2 Chronological sequence

The commencement of the negotiation with the JV-MC to commit to an MGP, as required under the terms of the original contract, is demonstrated by the client writing to JV-MC as early as 15 May 2008. This letter reminded the JV-MC of their obligations, having received the detailed design, to produce a priced bill of quantities and estimated maximum cost for the project within 30 days.

Throughout June, July and August the JV-MC was requested by both the client and engineer in the specially convened joint meeting of 13 July 2008 to submit the priced bill of quantities, but without result, the JV-MC stating they required more time.

In an effort to resolve this unacceptable delay and to expedite and assist the JV-MC, the client instructed the engineer to issue a copy of the Bill of Approximate Quantities produced by the engineer to the JV-MC which was transmitted on 26 August 2008 with the instruction that the JV-MC priced the Bill of Approximate Quantities and submitted within 30 days, or not later than 25 September 2008. The client subsequently allowed an extension to 15 October 2008.

On 15 October 2008 JV-MC submitted a lump sum price of US$959,747,820.50 excluding notably Stamp Tax, Income Tax and Customs Tax together with nineteen other qualifications to their price for the East Harbour complete and the West Harbour enabling works only. The JV-MC's price was unanimously rejected by both the client and the engineer, being double the preliminary estimate agreed at the signature of the contract. In addition, the duration of the contract and therefore the time for completion was revised to 1 September 2011.The JV-MC was instructed to revise and resubmit both the estimated maximum cost and the completion date for the project.

Despite continual requests and encouragement from both the client and the engineer, the JV-MC did not submit a revised estimated maximum cost until 15 December 2008. The proposal for $798,804,798 for the East Harbour complete and the West Harbour enabling works only excluded: the costs of extending the JV-MC bonds and guarantees, Stamp Tax, Corporate Income Tax, Customs Duties and fees. The completion date now announced was April 2011. While the decrease was a step in the right direction, the price was still considered excessive and the JV-MC was again instructed to revise and resubmit their estimate covering the whole scope of the works.

It took the JV-MC a further three months to produce a revised estimated maximum cost, submitting on 14 March 2009. The proposal for $773,986,609 was 'in accordance with the intent and terms of the contract, the prices do not include any taxes, or any price contingencies'. The completion date now submitted was March 2011. The proposal was presented and discussed at a meeting with the CEO of the client on 23 March 2009 and was rejected both in terms of price and time for completion with the CEO stating a final deadline for agreement under the cost-plus format being 30 April 2009.

The Board of the Engineer met the Board of JV-MC on 28 April 2009 in a final attempt to determine a fair and reasonable price and an agreed time for completion. At the end of the meeting, the JV-MC's price and terms and conditions were still wholly unacceptable at $769,700,000 compared to the engineer's estimate of $670,500,000.

On 17 May 2009 the parties met at the client's HQ and a lump sum price was presented by the JV-MC of $876,636,632 but again with substantial exclusions. In addition there were eleven further qualifications to the JV-MC proposal, with the revised date for completion announced as October 2011. This proposal was again rejected and the CEO requested JV-MC to submit a new price based on a bill of quantities and unit rates. A committee, comprising members of the client's and engineer's organisations, was assigned to negotiate unit rates with JV-MC. After the meeting held on 17 May 2009, 15 further meetings were held between the client, engineer and JV-MC up to the end of June 2009 to review all the components of the unit prices for concrete works. An agreement was achieved during the meeting held on 28 June 2009 with a unit rate for concrete, including formwork, rebar as well as indirect costs, and overheads and profits.

Following the letter issued by the client to the JV-MC on 18 June 2009 confirming the requirement for the contract to be amended, the draft Addendum was produced by the engineer and issued to the JV on 4 July for their review and comments.

The JV submitted their comments in sections, (as set out below) but took until 8 August 2009 to submit their comments to the final Article, a total of 35 days.

The client committee and engineer reviewed the comments on the Articles from the JV-MC as they were submitted throughout July 2009 and met with the JV for a three-day workshop on 2–4 August to review the 60 Articles.

Further, comments resulting from the three-day workshop were reviewed by the JV-MC on 8 August. The client committee and engineer again reviewed the 60 Articles and the Addendum, respecting local law and regulations, was issued on 10 August 2009. However, the majority of the 12 Annexes and 8 Attachments to complete the contract amendment were still outstanding.

The JV-MC agreed to submit the balance of the Annexes and Attachments by 19 August. They did not and a meeting was held on 20 August 2009 to determine the delay and establish a final date for submission. The JV-MC agreed to submit the majority of the outstanding documentation on 23 August 2009 but the bill of quantities would not be submitted until 31 August 2009.

The client and engineer reviewed the final documentation throughout September. The Completed Addendum with Annexes was issued on 1 October 2009 back to the JV-MC with the Attachments being issued the following week with the exception of those concerning cost.

The JV-MC were invited to submit their Price, to reflect Addendum Documentation as issued to them on 1 October, on Sunday 11 October with final negotiation to take place on 11 and 12 October 2009.

3.7.3 Estimate of 'cost-plus'

The engineer estimates the total cost of work completed to date on the cost-plus arrangement will be in the order of $84 million to $90 million.

3.7.4 Duration required for completion of the works

JV request for extension of time

On 30 November 2008, the JV-MC submitted a request for time extension claiming to postpone the contractual completion dates as follows:

- *East Harbour: from 1 September 2009 to 29 September 2011 (+ 758 days)*
- *West Harbour: from September 2009 to, 'at a minimum' 29 September 2011 plus a further time extension* to be claimed at a later stage depending on the change to scope of work to be agreed upon (this further time extension being taken as 6 months on an interim basis, i.e. end of March 2012).

The JV-MC put forward various factors it claims were beyond its control, part of them having already been assessed for their impact on the time schedule as of the cut-off date of the request for time extension (31 October 2008), the others being under development and for which the impact on the time schedule may lead to further claim once they are fully determined.

After an in-depth analysis of the JV's request, the engineer recommended:

- *East Harbour: 145 days + 62 days = 207 days*
- *West Harbour: 145 days + 134 days = 279 days*

3.7.5 Risk of delays

In the event that the client, for any reason, cannot deliver any one of the above, the contractor would be entitled to claim for an extension of time – and cost as a result. Therefore, the contract attachments must be very carefully considered as it directly implies a contractual obligation on the client.

For example:

1) Risk of delays beyond the control of JV-MC.
   a) Obstacles on site: estimated delay 2 months for clearing and allowing JV to access site area.
      i) East Harbour: fuel tanks (foundations), including underground networks
      ii) West Harbour: fuel pits and delay extending fence lines.
2) Delivery of essential utility services to JV-MC to start testing and commissioning.
   a) East Harbour: Estimated delay 4 months (indirectly caused by other packages outside the JV-MC Contract, failing to deliver services).
3) Therefore, based on the hypothetical assumption made above, the JV-MC would be delayed by a total of 6 months, and entitled to an extension to their contract period for completion. In doing so the contractor will remain on site for an additional period of 6 months.

Evaluation of the likely impact of the above, can be calculated in a very simple and broad-brush method, using the likely monthly cost shown below: (e.g. 6 months × $8 million per month = $48 million).

3.7.6 Evaluation of additional costs if completion of works extends beyond December 2011

The principal additional costs will be for indirect costs together with the extension of the insurances and letters of credit. This would apply to both JV-MC and principal subcontract packages, together with the engineer's time.

An estimated value for this would be $8 million per month, calculated as follows:

Letter of credit/insurances and guarantees	$0.80 million
Support facilities – main camp	$0.20 million
JV-MC indirect and general direct costs	$3.75 million
Subcontract work packages [indirect costs]	$2.05 million
Total cost for contractors	*$6.80 million*
Engineering costs	$1.20 million
Total cost per month	*$8.00 million*

3.7.7 Evaluation of the delay penalty to be applied under the current contract

In case the works are to be completed by December 2011 (i.e. the current estimation of the completion date), the total delay to be taken into account for the delay penalty would be:

End of December 2011 – 1 September 2009 = 28 months.

Out of these 28 months, 7 months (213 days) have to be discounted as an allowance for time extension (see discussion above), making a total of 21 months for delay.

These penalties should then be:

$150,000 × 21 months × 30 = $94,500,000

However, as per the provision of the contract, the delay penalties are capped by a ceiling of 5 per cent of the estimated maximum cost (EMC). According to whether we consider the EMC preliminary estimate of $485 million or the latest proposal of the JV-MC for $774 million, the maximum delay penalty would be:

a) For an EMC of $485 million, × 5 per cent = *$24.25 million (minimum estimate)*, or
b) For an EMC of $774 million, × 5 per cent = *$38.70 million (maximum estimate)*.

3.7.8 Analysis of JV cost proposal

Engineer's unit cost analysis of concrete works

As part of the ongoing contract negotiations, the engineer has undertaken numerous cost exercises and cost analyses. To date the main production on site has been the concrete works and therefore attributes the bulk, if not all, of the expenditure to date.

The engineer issued a detailed 'Unit Cost Analysis for Concrete Works' in July 2009. The use of cost analysed data is fundamental in negotiating the Addendum Contract Value.

- *Engineer's average unit cost, based on JV-MC BOQ submitted 23 June 2009 and 16 July 2009*:
  - engineer's unit summary rate
  - engineer's unit cost breakdown
  - average unit cost of concrete executed to 31 May 2009.
- *JV-MC average unit cost, based on JV-MC BOQ submitted 23 June 2009 and 16 July 2009*:
  - JV-MC unit summary rate
  - JV-MC unit cost breakdown.
- *Summary of quantities, based on JV-MC BOQ submitted 23 June 2009 and 16 July 2009*:
- *Engineer's unit/elemental cost of concrete for:*
  - columns
  - floor slab
  - ramps
  - walls
  - stairs
  - topping slab
  - piles
  - lean concrete
  - pile caps
  - tanks
  - beams
  - precast beams.

Engineer's analysis of indirect and support costs

As part of the detailed analysis carried out by the engineer, a detailed breakdown of JV-MC's indirect and support costs has been extracted from the submitted BOQ.

The details are summarised under the following headings:

- Site Management/Supervision/Administration Team
- General Expenses (schooling, housing, air tickets etc.)
- Mobilisation and Demobilisation Costs
- Communication and IT Costs
- Catering for Management Team
- Security Services
- Computer Hardware
- Specialist Advisers/Consultants
- Stationery
- Maintenance and Cleaning of Offices
- Site and Subcontractor's Support Costs.

The principle

Since 17 May 2009 agreement by the parties that the contract should be amended to a re-measurement basis, the unit prices were to be fixed with all indirect costs encompassed within. Indeed, a bill of quantities submitted unofficially on 16 July was presented in this manner. However, in the bill of quantities submitted officially on 3 September 2009, the indirect costs were submitted as a single line item to be divided into monthly payments over the agreed duration of the contract. The JV-MC support costs for the self-performed works were also presented as a line item entitled General Direct Cost.

In order to enable a proper comparison of the unit rates required by the client, the engineer has distributed the indirect and general direct costs, shown as lump sum line items, into the self-performed works and subcontract packages.

Overheads and profit

The original contract allowed overheads of 8.5 per cent together with 13.5 per cent for profit which were deemed appropriate at the outset of the project to enable the JV-MC to mobilise in the country and carry out the works in a very tight time schedule for the completion of the project.

However, with the amendment of the contract to re-measurable, the new financial environment and a less onerous time schedule, the profit margin of 13.5 per cent is no longer considered appropriate. The engineer therefore recommends that a figure of 10 per cent is allowed for self-performed works and 5 per cent where the works are performed by subcontractors. These rates are advised by the engineer as a compromise between the profit margin in the original contract and benchmarking from similar international projects.

Cost comparison

Following the redistribution of the indirect and general direct cost by the engineer, three separate cost comparisons were prepared applying the revised engineer's assessment of costs of work and overheads and profit.

- Cost comparisons 1:
  - Component 1: Minor evaluation differential.
  - Component 2: This clearly identifies that when all the indirect costs are added to the JV-MC price for self-performed works, especially in the concrete works, there is a 30 per cent differential between the JV-MC price and the engineer's estimate.
  - Component 3: This identifies the excessive support and indirect costs of the JV for the works carried out by subcontractors and includes the reduction in their profit.
- Cost comparison 2:
  - In this comparison the unofficial bill of quantities submitted by the JV-MC on 16 July 2009 was used that showed all the indirect costs included in the unit prices.
  - While not conclusive, as the contract price was different, the concrete works demonstrate the unit rates adjusted in Comparison 1 by engineer were reasonable.
- Cost comparison 3:
  - This comparison demonstrates the additional costs if the contract period is extended by six months which is a major concern of the client. A six-month overrun of the schedule contract duration has been estimated by the engineer to cost the client in excess of $22 million.

3.7.9 Benchmarking

Using the engineer's detailed knowledge of harbour works worldwide and adjusting the cost data for this project, a direct comparison can be made graphically of the cost shown as $/m^2.

The cost summary sheet and graph data have been 'normalised' to show cost on an equivalent basis. For example, the current submitted cost from JV-MC is for East and West Harbours, but West Harbour finished to basic marine work only. Therefore in order to give a comparison on a like-for-like basis, the engineer has estimated the likely cost of fully completing the West Harbour.

The cost data should be used as guidance only, but the indication within the current JV-MC submission is equal to $4,434/m^2 whereas the engineer's analysis is $3,678/m^2.

Using the data above and selecting the average cost per m^2, the engineer reinforces the estimate/analysis provided to date of **$779,690,931** for two *complete* Harbours, which for this analysis, and to allow a comparison of other similar works, includes the total cost of completion.

3.7.10 Costs incurred as a consequence of the claim/dispute with JV-MC

Costs as a result of termination

The existing contract provides for termination only on the grounds of default of the contractor. If the contract is terminated by the client, this is likely to result in a formal dispute and possible arbitration.

Termination for default – which could be for non-respect of the time for completion, initially scheduled for September 2009; the consequence for the contractor would be as follows:

a) site inventory carried out at the contractor's expenses
b) delay penalty, which is valued at $150,000 per day of delay, limited to 5 per cent of the estimated maximum cost
c) confiscation of the performance bond (10 per cent of estimated maximum cost) to cover eventual additional prejudices incurred by the client.

Furthermore, the client will be entitled to entrust the works to any other contractor of their choice at the expense (in case of cost overruns) and under the responsibility of the JV-MC.

The JV-MC would obviously challenge the claim that it is in default, alleging a number of extenuating circumstances which have been expounded in their interim claim for time extension as of 31 October 2008. It is obviously hard to predict the outcome of arbitration, but it would be safe to consider that blames might be laid – to various extents – on both parties.

In this case, the JV-MC would be entitled to payment, in addition to the works executed and materials/goods already ordered for the site, for the following items:

- any reasonable cost for removal from site of temporary facilities units, equipment, goods, materials and other instruments and related removal costs
- any other reasonable costs incurred or that would be incurred by the JV-MC as a result of [suspension or] termination, which are not reimbursable costs according to any other provision of the contract.

The first item would be the cost of dismantling (manpower), packaging of various equipment/goods and dispatching such items, at least from the site to the port of embarkation. This item can be roughly assessed to $15 million.

The second item would cover such costs as penalties for terminating the contractor's contracts with various suppliers/service providers (such as insurance, site surveillance, cleaning, etc.), as well as redeployment of their own team members on other projects. This kind of cost is more difficult to assess (redeployment costs) and may lead to dispute. A rough assessment of such costs could be around $5 million.

3.7.11 Costs of the project manager during the interim period

The termination of the JV-MC contract will require the engineer to carry out the following activities:

- carrying out an inventory of the works site and closing out the accounts (at least from the client's point of view, even if a dispute develops with the JV-MC)
- preparing the tender documents for the new contractor to be selected, with particular emphasis on quantifying the works executed and those to be executed
- assessing the bid(s) received from the new contractors
- assisting the client in negotiating the terms and conditions of the new contract and finalising all required contract documentation.

As already determined by the engineer, all these activities, as well as the mobilisation of the new contractor, are likely to take between 9 and 12 months, during which the services of the project manager will be required.

3.7.12 Costs for the taking over of the works by a new contractor

Costs to assure the custody of the works until a new contractor takes over the site

In case of termination, since the JV-MC will no longer be responsible for the site, a number of precautions will have to be taken before a new contractor takes over to make sure that:

a) the JV-MC vacates the site without delay
b) the site is properly secured and guarded to avoid theft of materials or equipment
c) precautions are taken to avoid damage during the suspension of the works (e.g. in case of inclement weather).

3.7.13 Costs of insurance of the works to be executed by the new contractor

If a new contractor is designated to complete the works, there will be a delicate issue to solve with regard to the insurance of the works, both during construction (for the remaining portion of the works to be executed) and after completion of the project.

During construction, the works are generally covered by a 'Contractor's All Risks' (CAR) insurance, valid from the start of the works until substantial completion and taking over by the client.

After the taking over by the client, and in accordance with the local laws, a contractor has a 10-year liability to compensate or make good any defect which may cause the partial or complete collapse of the works. This liability can be covered by a specific insurance policy.

In case of termination of the JV-MC, the new contractor will have to complete the works which have been started by another contractor, which will probably be considered by the insurance market as carrying greater risks than a project which has been started and completed by the same contractor.

Therefore, the issue of insurances will have to be carefully considered to identify suitable solutions, with specific terms and conditions, for insuring the works (during and after completion) in this particular context. Such issue should be borne in mind, even though at the present juncture it is not possible to evaluate a specific cost, both for conducting such study and for the possible ensuing insurance extra cost.

3.7.14 Costs for 'complexity factor'

Although very difficult to assess, it can be easily understood that the fact of taking over, half way, a site commenced two years earlier carries an inherent complexity due to such factors as the necessity to carry out a thorough due diligence of the site, to revisit the whole design and shop drawings of the project, to analyse all the tenders for the various works packages (already awarded or under evaluation/negotiation), etc.

Therefore this complexity factor will be reflected in the rates/prices of any new contractor to an amount which can be evaluated anywhere between 2 to 4 per cent of the present works estimate, i.e. in a range of $15 million to $30 million. The stage at which the works would be interrupted (before or after completion of the concrete works), will determine to a large extent the amount of this complexity factor.

3.7.15 Costs linked to the present context of the international markets

According to current markets trends, the tender costs are likely to be very competitive and, given the slight rise in raw material costs, tender costs have remained static or fallen. Therefore retendering is unlikely to result in significant increase in construction costs.

3.7.16 Costs of delayed availability of the facility (loss of profit)

As already noted, terminating the JV-MC contract is likely to entail a 9- to 12-month delay on the total project completion. Such delay will lead to a loss of revenue for the client and other concessions within the facility, from mooring fees, rentals of shops, restaurants and other commercial ventures in or around the harbour.

Concerning other charges, a new harbour of this size fitted with state-of-art equipment and facilities is likely to generate a doubling of the present level of such charges. Considering the usage as it presently stands, the postponement of the commissioning of the new harbour by one year might thus induce a loss of charges between $11 million and $15 million.

With regard to non-core charges, given their level as of to date, they have to be considered almost as a new revenue stream. When the new harbour starts operation, considering that in modern facilities such charges may equal or exceed harbour fees, an additional loss of revenue between $22 million and $30 million is a reasonable estimate of lost revenue by the postponement of the commissioning of the new harbour by one year.

The overall impact of the postponement of one year, to be expected in the event that the JV-MC is terminated, can therefore be roughly estimated to $45 million.

3.7.17 Discussion/tutorial questions related to this case study

1 Was the original decision to award the contract on a lump sum basis correct? Justify your answer.
2 What were the major financial risks to the client?
3 What was the purpose of the benchmarking exercise?
4 Outline the various financial 'complexities' which created a difficulty in moving from cost-plus to a measurement contract.

4 The stages of development of construction projects

4.1 Introduction

Construction projects start with an idea in the client's mind about what assets they would like to have at the end of the process and finish off with the actual finished building, which hopefully satisfies that original idea. If it doesn't, then something has gone wrong somewhere, but that is not for this book to discuss. Therefore, the project progresses through several stages:

a) inception of the project and preparation for its delivery, including setting an overall budget (i.e. what can the client afford?)
b) outline sketch designs to show the client various alternatives for how their ideas can be built
c) a development of the chosen design from these rough sketches to concept designs where the 'big decisions', such as choices of construction (what type of foundations, how many floors, steel frame, concrete frame, load-bearing walls etc.) will be made
d) further development of the design with more detailed decisions on materials, finishes etc., including alternatives for the client to choose
e) final design with drawings and specification at a level of detail suitable for actual construction
f) construction of the project on site
g) commissioning, handover and completion.

As shown in Figure 4.1, there is a decision point at the end of each stage, where the client will say 'Yes, that's fine – that's still what I want – carry on guys'. In order to be able to do this, the client will not only wish to see the development of the design solution, but will also wish to see the anticipated cost of that solution to ensure that the cost remains within their budget. This anticipated cost must be regularly updated as critical design decisions are made and incorporated into the project. To give an example with which we are all familiar, we would all like to have the best possible house for our needs, but unfortunately, most of us have to live within a budget and the bank will only lend us a certain amount of money, so the decisions are invariably constrained by what funds we have available at the time.

Therefore, there needs to be a reasonably accurate cost estimate produced at each of these stages, to enable the client to progress with the works or make economies somewhere to reduce the costs, or as a last resort, to cancel the whole thing if the costs

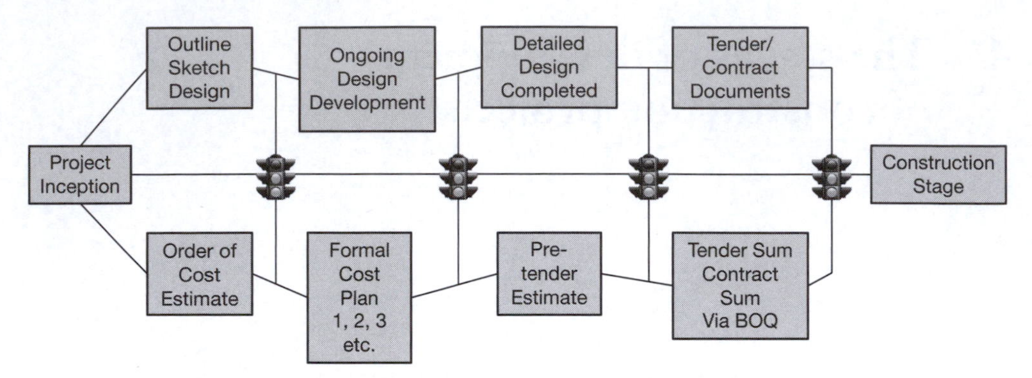

Figure 4.1 Stages of the design process with cost estimates and decision points

look like they have spiralled out of control. The required levels of accuracy of these cost estimates will vary at the different stages of the project, ranging from rough estimates or 'ball-park' figures in the early stage to much more reliable and accurate estimates (or quotations – see discussion at beginning of Chapter 1) as the project gets closer to the construction stage. See also Figure 5.3 in Chapter 5. As design decisions made during the early stages of a project life cycle are more global and tentative than those made at a later stage, the cost estimates made at the earlier stage cannot be as accurate and, of course, the accuracy of a cost estimate will always be a reflection of the quality and detail of information available at the time of estimation. See Figure 4.2 for an illustration of these techniques as the design stage of the project progresses.

As mentioned above and in previous chapters, cost estimates can be classified into several major categories according to how they are used at the different stages of the project. A construction cost estimate can therefore be used:

a) to establish an initial 'ball-park' figure (feasibility estimate)
b) to serve as a 'reality check' on the development of the design (design estimate or cost plan)
c) to benchmark the tender bids when they are received from contractors (bid estimate)
d) to build up the real costs of the works by the proposed contractor for conversion to a tender figure and contract price (tender estimate)
e) to control the construction stage expenditure, especially variations (control estimate).

Consequently, the different stages of the project will use cost estimates for different purposes, and we shall see in subsequent chapters how these estimates are slightly different from each other and are produced using different techniques. At present, it is worth noting that the types (and accuracy) of cost estimates reflect the stage of design that the project has reached.

Table 4.1 itemises the stages in the life cycle of a construction project using the principles described above. The two main classifications used in the UK industry are the *RIBA Plan of Work* (Column A) and the *OGC Gateway* (Office of Government Commerce) in Column C. Although each model names the stages slightly differently,

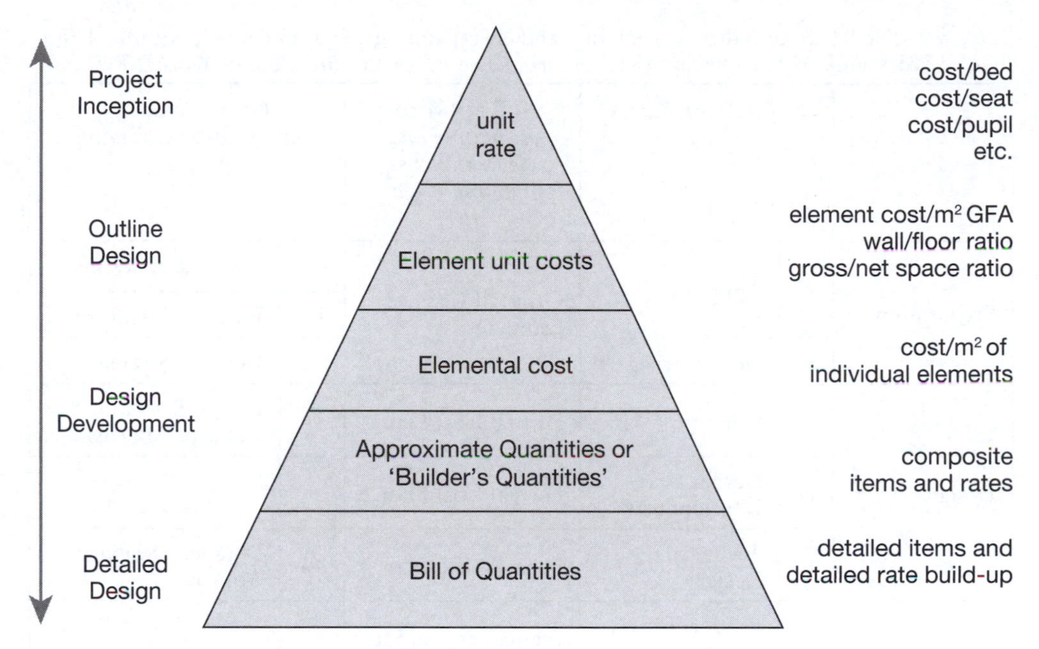

Figure 4.2 Hierarchy of estimating data

the principles are broadly the same, from the decision by the client at the outset regarding why they need the project, right through to taking over the finished article and working out whether they have actually got what they originally wanted (which is always a scary moment, as anyone knows who has bought a product which looked good in a shop but doesn't quite have the same gloss in the cold light of day at home).

Let us now look in some detail at the two models, in terms of cost estimates and the helpfulness of the model for early contractor involvement (ECI).

4.2 RIBA Plan of Work

The RIBA Plan of Work describes the activities required in a construction project starting from the beginning of the process, when the client's requirements are appraised by the initial project team, through to the post-construction stage when the facility is in operation by the end-user. When the Plan of Work was originally developed, it was designed very much with the traditional procurement route in mind, so the architect was in charge, other consultants made their subservient contributions and the contractor was kept very much in their place by not being allowed to contribute to the design (and the general feeling was 'why should they contribute? What possible input could they have at the design stage – that's the architect's job').

Fortunately, considerable progress in attitudes has been made since the Plan of Work was first published and although the Plan of Work stages A–L are still fundamentally the same, the descriptions of the key tasks within each stage together with the list of possible contributors to the stage, has been updated to make appropriate allowance for the early involvement of the contractor.

Table 4.2 illustrates the content and requirements of each stage, together with the format of cost estimate which would be appropriate at each RIBA Plan of Work stage.

Table 4.1 The RICS formal cost estimating and cost planning stages in context with the RIBA Plan of Work and OGC Gateways (adapted from *The RIBA Outline Plan of Work* 2007).

	RIBA Work Stages (A)		*RICS NRM cost estimating and elemental cost planning stages (B)*	*OGC Gateways (applicable to building projects) (C)*	
Preparation	A	Appraisal	Order of Cost Estimate	0	Strategic Assessment
				1	Business Justification
	B	Design Brief		2	Delivery Strategy
Design	C	Concept	Formal Cost Plan 1	3A	Design Brief and Concept Approval
	D	Design Development	Formal Cost Plan 2		
	E	Technical Design	Formal Cost Plan 3	3B	Detailed Design Approval
Pre-construction	F	Production Information	Pre-tender estimate		
	G	Tender Documentation			
	H	Tender Action	Post-tender estimate		
Construction	J	Mobilisation		3C	Investment Decision
	K	Construction to Practical Completion		4	Readiness for Service
Use	L	Post-practical Completion		5	Operations Review and Benefits Realisation

Source: The NRM published by the RICS.

4.3 OGC Gateway

The OGC (Office of Government Commerce – part of the UK government) Gateway Process is designed to examine programmes and projects at several key decision points in their life cycle. The examination should be carried out at the beginning of each stage of a project to look ahead in order to provide assurance that the project can progress successfully to the next stage. The Gateway Process is part of best practice in UK central civil government (a mandatory requirement), the health sector, local and regional government as well as defence projects.

According to the OGC website, Gateway Reviews are not just concerned with construction projects, but are equally applicable to a wide range of programmes and projects including:

Table 4.2 The RIBA Plan of Work stages and format of cost estimates

STAGE OF RIBA PLAN OF WORK	*ESTIMATE REQUIRED*
STAGE A – APPRAISAL **Key tasks** – Identification of client's needs and objectives and preparation of both technical and financial feasibility studies **People involved** – Traditionally, the client team and lead consultant(s) would agree the key tasks above. Usually a design consultant (architect) and cost consultant (QS). **What happens with ECI?** – If the contractor is involved at this stage, this would represent a 'turnkey' project where the contractor is given total responsibility and authority for the project from beginning to end. Just hand over the keys at the end, please.	Ball-park estimate required to assess feasibility of scheme by a comparison with other similar schemes.
STAGE B – DESIGN BRIEF **Key tasks** – Establishing and refining client's requirements and assessment of most suitable procurement method. Design would be very basic sketches showing overall area and height (i.e. number of storeys). Different design proposals would be available to the client, which are all suitable for their requirements. **People involved** – Traditionally, again just the client team and lead consultant(s). **What happens with ECI?** – If the contractor is involved at this stage, the procurement route has already been chosen in principle. The contractor would be able to offer some strategic advice on the buildability of the different sketch designs.	Refined ball-park estimate based on costs per square metre of similar construction, updated to the proposed construction start date and location of the works.
STAGE C – CONCEPT DESIGN **Key tasks** – Providing the client with an appraisal (both technical and financial) of the different proposals, so they may determine the form of the project. Develop the design by making the 'big' decisions – i.e. what type of foundations? What type of load-bearing superstructure? **People involved** – Client team and design team, which by now may include architects, structural engineers, services engineers, quantity surveyors and planning consultants. **What happens with ECI?** – If the contractor is appointed on a two-stage tender, or under partnering arrangements such as PFI, the preliminary stage would be commenced, possibly up to preferred bidder stage.	Estimate further refined based on cost per square metre, although information on the separate building elements may now become available, which refines the estimate further using the circumstances of this project, rather than separate but similar projects.
STAGE D – DESIGN DEVELOPMENT **Key tasks** – The brief should not be modified after this point (Yes, I know – dream on, but the client should be made aware that changes to their requirements after this point will result in abortive work by the design team, which the client will ultimately pay for). This stage is where the concept design is developed in earnest with the overall sketch designs being 'fleshed out' and the specification of the individual elements becomes more definite.	The project estimate is subject to continuous refinement through the procedures of cost planning. Each significant design decision is put into the cost plan to ensure the overall estimate is still realistic. Value engineering proposals are compared with each

(Continued)

Table 4.2 (Continued)

STAGE OF RIBA PLAN OF WORK	*ESTIMATE REQUIRED*
Critically, this is when all the elements of the design (architectural, structural engineering and services) must come together in a coordinated and complete package. Full planning applications are made at this stage. **People involved** – Full design team and any statutory authorities. **What happens with ECI?** – This is the stage where early the appointment of the contractor has the most benefit. The contractor's contribution to buildability and value engineering can be significant if used properly.	other to ensure the most economical solution is chosen. Increases in one element may be offset by savings in another.
STAGE E – TECHNICAL DESIGN/FINAL PROPOSALS **Key tasks** – Completion of the client's brief, planning and other statutory approvals should have been obtained. At completion of this stage, the design of the project should be finalised and in a form which can be built. **People involved** – All members of the design team. **What happens with ECI?** – The contractor may have already mobilised and commenced some preliminary works by this stage. This is another major advantage of ECI, that the contractor can start some 'enabling works' e.g. site set-up, site clearance, preliminary excavation etc. before the full design has been completed, thus saving time and costs for the project.	Final cost plan, giving the estimated cost of the project in elemental form. All cost data is project specific with all comparison costs having been eliminated.
STAGE F – PRODUCTION INFORMATION **Key tasks** – Final decision 'signed off' on all matters relating to design, specification and estimated cost. For a traditionally procured project, tender documents are prepared allowing the rendering contractors to accurately price the works. **People involved** – All members of the design team, plus contractor if appropriate. **What happens with ECI?** – If they have already been appointed, the contractor would be heavily involved in this stage and the production documents would also include the method statement and programme/schedule. The contractor will have already mobilised and commenced the preliminary works.	Final pre-construction estimate. Under ECI, the estimate at this stage would have been agreed by the contractor as the contract price, depending on when the contract point occurred in the negotiations.
STAGE G – TENDER DOCUMENTATION **Key tasks** – This stage occurs at this point only in the traditional procurement route, i.e. when contractors are not involved in the design stage and are appointed solely in price competition on a completed design. The tasks are therefore to prepare and collate tender documents to enable tenders to be obtained as an accurate lump sum. **People involved** – The full design team to prepare tender drawings and specifications, together with cost consultants/ QS to prepare pricing documents/BOQs. **What happens with ECI?** – If they have already been appointed, this stage would already have occurred and the tender documents would have been quite different because no firm design information was available.	Final bid estimate by the cost consultant, who holds their breath in hoping that the bid estimate (prepared using historical costs) is close to the actual tender (prepared using actual resource costs).

STAGE OF RIBA PLAN OF WORK	*ESTIMATE REQUIRED*
STAGE H – TENDER ACTION **Key tasks** – Prepare and complete tender documents, develop list of tendering contractors, evaluate tenders following submission and appoint the successful tenderer. **People involved** – Mainly the tendering contractors to calculate the bid price (see Chapter 6) although the design consultants and contract/cost consultants will be available to respond to queries on the tender documents submitted as RFIs (requests for information). The contract/cost consultants are normally responsible for evaluating the tender submissions and recommending appointment. **What happens with ECI?** – If they have already been appointed, this stage would already have occurred. However, a final contract price still needs to be agreed at design completion so that the client still has some cost certainty in the project.	The 'estimate' is now calculated quite differently. The cost consultants' 'bid estimate' still exists, but the contractor produces their own estimate from first principles (i.e. resource costs). This is converted to a tender sum (see section 7.2) and then to a contract sum (see section 7.3)
STAGE J – MOBILISATION **Key tasks** – Site handed over to the contractor to enable start of work. **People involved** – All parties. **What happens with ECI?** – If they have already been appointed, this stage would already have occurred.	Not applicable
STAGE K – CONSTRUCTION **Key tasks** – Operations on the site. Construction of the works from commencement to practical/substantial completion including commissioning. **People involved** – All parties. **What happens with ECI?** – If they have already been appointed, this stage will overlap considerably with the design stage, thus reducing the overall duration of the project. Design information is transmitted to the contractor as required for construction and normally in accordance with a design information release schedule (IRS).	The contract sum/contract price becomes the budget control figure, replacing the pre-construction estimate. During this stage, estimates are required for variations, changes, additions to the scope of works (see Chapter 8)
STAGE L – POST-PRACTICAL/SUBSTANTIAL COMPLETION **Key tasks** – Contractor responsible for completing remaining minor works and snags. Snag list/punch list produced by supervision consultants. Final inspection prior to final handover. Preparation of final account. **People involved** – All parties especially contractor and supervision consultants (usually designers). Cost consultants and contractor to agree final account. **What happens with ECI?** – No difference. The contractor may be appointed early in the design process but they are still required to be there to the end of the project.	The final account is not an estimate. This is what the project actually cost and it is salutary to compare the actual final cost with the estimate(s) produced during the design stage. It is also very revealing to compare the final account with the tender figures produced by all tenderers.

- policy development and implementation
- organisational change and other change initiatives
- acquisition programmes and projects
- property/construction developments
- IT-enabled business change
- procurements using or establishing framework arrangements.

Underlying the OGC Gateway process is a set of guiding principles which define what the Gateway is trying to achieve. These principles are applied by all Gateway users in order to maintain brand standards.

It is not the purpose of this book to discuss the Gateway decision points in detail; however, Table 4.3 illustrates the content and requirements of each stage, together with the format of cost estimate which would be appropriate at each stage.

Table 4.3 The OGC Gateway project stages and format of cost estimates

STAGE OF OGC GATEWAY	*ESTIMATE REQUIRED*
REVIEW STAGE 0 – STRATEGIC ASSESSMENT This stage is mainly a review of the 'programme' (which may be a series of individual projects) and is essentially a review of the outcomes and objectives for the programme (and the way they fit together) to confirm that they make the necessary contribution to the overall strategy of the organisation. The review is also required to: • Ensure that the programme is supported by key stakeholders. • Confirm that the likely success in achieving the objectives has been considered in the wider context of client policy and procurement objectives, together with the organisation's delivery plans. • Assess the arrangements for leading, managing and monitoring the programme as a whole and the individual projects. • Review the arrangements for identifying and managing the main programme risks (and the individual project risks). • Check that provision for financial and other resources has been made for the programme (initially identified at programme initiation and committed later) and that plans for the work to be done through to the next stage are realistic, properly resourced with sufficient people of appropriate experience and authorisation. • After the initial review, check progress against plans and the expected achievement of outcomes. • Check that there is engagement with the market as appropriate on the feasibility of achieving the required outcome.	As this is a strategic assessment of the project or programme of projects, there will be no individual project estimates at this stage. The client is required to ensure that the financing structure is in place. The actual figures and budgets will be established in a later stage.
REVIEW STAGE 1 – BUSINESS JUSTIFICATION This stage is about confirming that the business case for the project is robust and that it meets the business needs and is affordable, achievable, and appropriate options have been explored in order to achieve value for money. The stage should also: • Confirm that appropriate expert advice has been obtained as necessary to identify and/or analyse potential procurement options.	Order of cost estimate (see section 5.1.2).

STAGE OF OGC GATEWAY	*ESTIMATE REQUIRED*
• Establish that the feasibility study has been completed satisfactorily and that there is a preferred way forward, developed in dialogue with the market where appropriate. • Confirm that the market's likely interest has been considered. • Ensure that there is internal and external authority, if required, and support for the project. • Ensure that the major risks have been identified and outline risk management plans have been developed. • Establish that the project is likely to deliver its business goals. • Confirm that the scope and specifications are realistic, clear and unambiguous. • Ensure that the full scale, intended outcomes, timescales and impact of relevant external issues have been considered. • Ensure that the desired benefits have been clearly identified at a high level, together with measures of success and a measurement approach. • Ensure that there are plans for the next stage. • Confirm planning assumptions and that the project team can deliver the next stage. • Confirm that overarching and internal business and technical strategies have been taken into account. • Establish that quality plans for the project and its deliverables are in place. • Confirm that the project is still aligned with the objectives and deliverables of the business strategy to which it contributes.	
REVIEW STAGE 2 – DELIVERY/PROCUREMENT STRATEGY At this stage, the outline business case should be fully defined. At this stage, it is also necessary to: • Confirm that the objectives and desired outcomes of the project are still aligned with the programme to which it contributes. • Ensure that the delivery/procurement strategy is robust and appropriate. • Ensure that the project's plan through to completion is appropriately detailed and realistic, including any contract management strategy. • Ensure that the project controls and organisation are defined, financial controls are in place and the resources are available. • Confirm funding availability for the whole project. • Confirm that the project will facilitate good client/supplier relationships in accordance with government initiatives such as Achieving Excellence in Construction. • Confirm that there are plans for risk management, issue management (business and technical) and that these plans will be shared with the contractor's supply chain. • For construction projects, confirm compliance with health and safety and sustainability requirements. • Confirm that the stakeholders support the project and are committed to its success.	Order of cost estimate still valid at this stage.

(Continued)

Table 4.3 (Continued)

STAGE OF OGC GATEWAY	*ESTIMATE REQUIRED*
REVIEW STAGE 3 – INVESTMENT DECISION As the OGC Gateway process is used for projects other than construction, this stage covers the decisions on the scope of works and the preparation of any tender documents in order to choose the 'suppliers' of the product. In construction projects, this covers what are arguably the most crucial stages of conceptual design, design development and tender. In construction, this stage has therefore been split into 3A, 3B and 3C as detailed in Table 4.1.	
3A – Design brief and concept approval • Confirm the full business case and project benefits plan. • Confirm that the objectives and desired outputs of the project are still aligned with the programme to which it contributes and/or the wider organisation's business strategy. • Check that all the necessary statutory and procedural requirements are followed. • Confirm that the recommended contract decision, if properly executed within a legal agreement is likely to deliver the outcomes on time, within budget and provide value for money. • Ensure that management controls are in place to manage the project through to completion, including contract management aspects. • Ensure there is continuing support for the project. • Confirm that the development and implementation plans of both the client and the supplier or partner are sound and achievable. • Confirm that there are plans for risk management, issue management and change management (technical and business), and that these plans are shared with suppliers and/or delivery partners. • Confirm that the technical implications, such as 'buildability' have been addressed.	Estimate based on concept design as stages C and D of RIBA Plan of Work (see Table 4.2 above).
3B – Detailed design approval • Ensure there is continuing support for the project. • Confirm that the final design to be approved can answer the points above. • Confirm that the objectives and desired outputs of the project are still aligned with the programme to which it contributes and/or the wider organisation's business strategy. • Confirm that the development and implementation plans of both the client and the supplier or partner remain sound and achievable.	Estimate based on final proposals as stage E of Table 4.2 above.
3C – Final go-ahead • Ensure there is continuing support for the project. • Confirm that the objectives and desired outputs of the project are still aligned with the programme to which it contributes and/or the wider organisation's business strategy. • Confirm that the development and implementation plans of both the client and the supplier or partner remain sound and achievable.	Estimate based on final proposals as stage E of Table 4.2 above.

STAGE OF OGC GATEWAY	*ESTIMATE REQUIRED*
REVIEW STAGE 4 – READINESS FOR SERVICE This stage is at the end of construction and the facility is almost ready for occupation by the client or other end-user.	Final account
REVIEW STAGE 5 – OPERATIONS REVIEW AND BENEFITS REALISATION This is the 'post-occupancy evaluation' stage when the benefits of the project should have manifested to the client.	Not applicable

4.4 The RICS New Rules of Measurement (NRM)

The New Rules of Measurement (NRM) are a suite of documents issued by the RICS Quantity Surveying and Construction Professional Group. These have been developed to provide a standard set of measurement rules that are understandable to all those involved in a construction project. The rules aim to take a cradle-to-grave approach to the procurement of construction projects, and include cost estimating, works procurement and post-construction procurement.

As well as covering the traditional costs that are reflected in measurable building work, the NRM encompass a range of other issues. These include overheads, profit and inflation, and other costs including consultants' fees, land costs, and planning obligations. Guidance is also provided on dealing with tax allowances and grants.

Prior to the New Rules of Measurement being published, the measurement of construction work was generally covered by the various standard methods of measurement (SMMs) available for the different sectors of the construction industry. The SMMs in common use were:

a) Standard Method of Measurement of Building Work (with the latest being the 7th edition – SMM7)
b) Civil Engineering Standard Method of Measurement (with the latest being the 4th edition – CESMM4)
c) Method of Measurement for Roads and Bridges
d) Principles of Measurement International – POMI.

Many different countries also have their own standard methods of measurement, which have been designed for their own particular procurement circumstances and legal jurisdictions.

The purpose of a standard method of measurement is to convey to the tendering contractor what is and what is not included in the measured items. They will then be able to include all appropriate costs in the rate for the item. All other costs should be included either in the preliminaries section or the overheads and profit spread percentage. This is a way of ensuring that the price build-up is as efficient and effective as possible.

Due to the wide spread of procurement strategies being used in the modern construction industry, it was felt that the rules of measurement needed to encompass wider project issues and be able to be used by the client organisation at the early stages of the project. Therefore, part of the rationale for the new rules is to provide central governments, local governments and other public sector bodies with a value-for-money framework. More accurate cost estimation should also give reassurance to banks and other financial institutions which are lending for construction projects. The relevant professional bodies) such as the RICS and RIBA clearly support the need to develop standard rules of measurement and reporting procedures for sustainability across the property, land and construction sectors through the use of documents such as the NRM. Although the NRM is based predominantly on UK practice, the basic principles should be applicable internationally.

The NRM will eventually comprise three major documents:

1 Order of Cost Estimating and Elemental Cost Planning
2 Procurement (an alternative to SMM7)
3 Whole Life Costing.

At the time of writing (August 2012), only the Order of Cost Estimating and Elemental Cost Planning has been published in its final form. The NRM for Procurement is in final draft form and Whole Life Costing is in preparation.

4.4.1 NRM – Order of Cost Estimating and Elemental Cost Planning

The document is structured in four parts.

Part 1 – General

This section introduces the purpose of the document and sets out how the NRM relates to the RIBA Plan of Work and OGC Gateway Process.

Part 2 – Measurement rules for the order of cost estimating

As we have seen previously, the order of cost estimate is the first 'ball-park' or feasibility estimate produced for a project, and up to now there has not been any industry-wide agreement of how this estimate should be structured or produced. It has been left to individual clients or professional consultants to structure the estimate in their preferred way, depending on their corporate requirements and in-house skills. Clearly, the feasibility stage is an important decision point in a project and any client rejection at this early stage will mean that the project does not progress further. However, since the 1970s, the Building Cost Information Service (BCIS), which is a subscription service and allied to the RICS, has developed an information service for its subscribers which gives considerable amounts of building cost information, for both construction costs and tender prices. The format used by the BCIS for these elemental costs has been taken to be the 'industry standard'.

Part 2 of the NRM is structured into the following sections, which illustrates how an order of cost estimate could be structured:

- The purpose of an order of cost estimate
- Information requirements for order of cost estimates
- Constituents of an order of cost estimate
- Measurement rules for building work
- Elemental method

- Unit rates and element unit rates (EURs) used to estimate the cost of building work
- Updating unit rates and other costs to current estimate base date
- Cost estimate for main contractor's preliminaries
- Cost estimate for main contractor's overheads and profit
- Project/design team fees
- Other development/project costs
- Risk allowances
- Inflation
- Value Added Tax (VAT) assessment
- Reporting of order of cost estimates.

As it is important that the receiver of information knows exactly what is and what is not included in the information or data received, these rules are vital to avoid confusion and to cut down any risk premiums associated with poor quality or ambiguous information.

Part 3 – Measurement rules for elemental cost planning

The measurement rules for what is the next stage in the design cost control process, i.e. the cost planning stage, include many of the items listed above together with rules for measuring building works which would normally be found in one of the standard methods of measurement. The purpose of the cost planning stage is to update the order of cost estimate as the design is being developed; therefore the structure of the estimate remains essentially the same.

Part 4 – Tabulated rules of measurement for elemental cost planning

This section contains the new information provided by the NRM. Up to this point, the rules of measurement have related only to the tender stage and the production of firm bills of quantities after a full design has been completed. The various SMMs contained very detailed instructions on how to measure building works (up to SMM edition 5, there was even an obligation to measure end, angles and intersections on skirting boards!), therefore there was little point in producing detailed bills of quantities unless the design had been completed.

The NRM for the cost planning stage now brings in the same type of measurement rigour to the upstream stages of design development. In section 4.5 of the NRM, the measurement rules are structured into the following 15 group elements:

1 Substructure
2 Superstructure
3 Internal finishes
4 Fittings, furnishings and equipment
5 Services
6 Complete buildings and building units
7 Work to existing buildings
8 External works
9 Facilitating works

10 Main contractor's preliminaries
11 Main contractor's overheads and profit
12 Project/design team fees
13 Other development/project costs
14 Risks
15 Inflation.

By comparing this list with the full list of building elements in Table 5.5 in Chapter 5, it can be seen how the cost plan is intended to be developed into a more detailed elemental cost estimate by the end of the design stage when all design decisions will have (hopefully) been taken.

4.5 Summary and tutorial questions

4.5.1 Summary

Cost estimates produced for a construction project should follow the standard pre-construction project stages as outlined in section 4.1. As each design stage is completed and submitted to the client or their representative for approval and/or acceptance, the design must be accompanied by an appropriate cost estimate so that the client can make their decisions on the three major project criteria – time, cost and quality. The two major models which are used for structuring the stages that a capital project must go through are the RIBA Plan of Work and the OGC Gateway process. Using these two models, the RICS NRM has established the types of estimates that are appropriate at each stage – see Figure 4.2 and Table 4.1.

The techniques used for developing and producing these estimates are discussed in detail in Chapters 5 and 6.

4.5.2 Tutorial questions

1 Why is it important to understand that all projects go through the same stages of development?
2 What are the main differences between the RIBA Plan of Work and the OGC Gateway process?
3 How much different would the stages look if the contractor/builder was appointed:
   a) at the design brief stage (RIBA Plan of Work stage B)
   b) at the technical design stage (RIBA Plan of Work stage E)
4 What is the major estimating technique used in each of the types of estimate given in Figure 4.2?
5 Comment on the major reasons why the RICS New Rules of Measurement (NRM) now include measurement rules for the cost estimating stage in addition to measurement rules for completed design.
6 Discuss the different functions of the various cost estimates during early contractor involvement in a project.
7 Was it worth appointing the lowest tenderer and getting all that grief when the final account turned out to be closer to the second/third lowest tender?

5 Estimating techniques at inception and design stages

As stated in Chapter 1, the early stages of the pre-contract phase of a construction project use 'approximate estimating' techniques to establish the likely cost, leaving resource-based estimating techniques to the design completion stage (Chapter 6). The preparation and early design stages (i.e. RIBA Plan of Work stages A to D in Table 4.1) only contain outline design information; therefore the cost estimate is equally 'outline' in character. As the design progresses in detail, so does the project estimate until a stage is reached where the detailed design is complete and approved by the client, allowing a final cost plan or pre-tender estimate to be developed.

5.1 Business justification and order of cost estimate phase

The business justification or preparation stage (see Table 4.1) is the initial stage of a project's life and consists of the selection of alternative proposals which should all meet the client's requirements of the completed project. Before a particular solution is selected, the client and its advisers must examine the investment value of all the alternatives available and make an assessment of the impact on the current business. This is clearly outside the scope of this book, but this 'development equation' is relatively straightforward in principle; on the one hand, there are the following development costs:

- land acquisition costs
- possible relocation costs of existing departments or tenants
- environmental compliance and mitigation costs, e.g. contaminated land remediation
- costs of obtaining planning and other statutory approvals
- design costs
- legal fees
- other consultants' costs
- construction costs
- contingencies.

The assignment of contingencies is very important at the business justification stage of the project as they are necessary to ensure that any unforeseen items of work or level of detail which may come to light later will not jeopardise the project and therefore give it a better chance to progress to the next stage. Contingency values may vary between various alternatives and the actual amount will depend on any significant risk differences.

On the other hand, the development benefits should outweigh the costs if the project is to be considered viable. These benefits would include any extra sales revenue generated by the increased capacity due to the project, together with any efficiencies gained by reducing operating costs (because of more modern equipment or relocation) and so on. A key focus should be, as with any corporate investment, the alignment of the project with the client's business goals and objectives.

Once these calculations have been completed and approved by the client's decision makers, the proposed project may then be formally approved and initiated. Approval and initiation of the project signifies that formal project activities can begin. All parties should continually remind themselves that a project's primary purpose is to meet the stated business goals and objectives of the client and before any formal project planning activities can occur, the client's internal project sponsor must approve the business outcomes at that specific point during project delivery. Approval indicates that the client agrees that there should be further investment in delivery of the project.

5.1.1 Key questions

Key questions that must be addressed during the business justification stage include:

- What business problem or issue does the project solve?
- What other alternatives have been considered?
- What is the impact of not doing this project?
- What are the expected benefits of the project?
- When will the project deliver the expected benefits and business outcomes?
- What are the risks of time and cost overrun?

5.1.2 Order of cost estimates

As mentioned above, during the business justification stage, an estimate for construction costs will be required for input into the development equation. This is where the crystal ball comes in handy, as there is no design yet, therefore there are no drawings, specifications or any of the other documents normally required to assess the extent of work and consequently build up an estimate. What is therefore required is a 'ball-park' figure based on previous similar projects, hopefully as recent as possible and with a location as close as possible to that of the intended project.

Additionally, at this feasibility study stage, various alternative solutions will be sketched out to the client for the building or facility and a preliminary construction cost estimate will need to be prepared for each of those alternatives, in order to assess whether each alternative solution will be able to be constructed for the budget available in the development equation.

Because the estimates are based on previously completed similar projects, the figures that will be used will often be the *tender prices* and the client is being given an estimate of what they are likely to pay for the works. The tender price to be used (which may or may not be the contract price/contract sum of a project) will normally be the lowest tender received for the project being considered and before any post-tender, pre-contract negotiations which often and not surprisingly results in an even lower contract price for the same scope of work. This will certainly be the case if the estimate is based on cost information taken from the public domain, i.e. from publicly available but subscription cost databases such as BCIS – Building Cost Information Service or from the various published price books, such as Spon's. However, many specialist firms of cost consultants maintain their own detailed cost breakdowns of previous projects and would therefore be able to more accurately determine anticipated costs of future projects, perhaps even using contract prices instead of tender prices.

This process can also be known as parametric estimating, which is defined as the process of using various factors to develop an estimate. The factors are based on engineering parameters, developed from historical databases, construction practices and engineering/construction technology. Parameters may include physical properties that describe the project (e.g. building size, building type, foundation type, external envelope material, roof type and material, number of floors, functional space requirements, interior utility system requirements, etc.). The appropriateness of selecting the parametric method depends upon the extent of project definition available at the time, the similarity between the project and other historical data models, the ability to calculate details, and known parameters or factors for the project.

What happens with ECI?

This 'order of cost estimate' is usually calculated and given to the client by a firm of quantity surveyors or cost consultants, who would be appointed by the client for that purpose, and possibly be retained by the client for cost advice throughout the design (or pre-contract) stage. However, most large contracting organisations are quite capable of giving this cost advice themselves and, strangely enough, to be objective in doing so. Such cost advice from a construction firm would be especially beneficial if the company had constructed similar projects in the recent past which resulted in positive client satisfaction. Therefore, clients should not be afraid of employing construction firms on a 'package deal' or design-build basis.

Example

A client wishes to build an office block in Nottingham, UK, in order to relocate several departments of the organisation from the Greater London area, thereby reducing operating costs. The client requires medium density cellular office accommodation for 250 staff with appropriate circulation space, stairs, lifts, toilets, kitchens etc.) to be available in 24 months. How much can the client expect to pay for construction work?

Well, that's a fairly straightforward question, so how should the 'order of cost estimate' be developed?

First, a ball-park figure can be calculated using functional unit costs. Table 5.1 is taken from *Spon's Architects' and Builders' Price Book* 2012. Using this data, a medium density air conditioned office building would require approximately 15 m^2/person at an average rate of £26,500 per person (at 2012 national average prices). Different parts of the UK have different tender levels, depending on the relative competitiveness in the various regions and also the costs of labour and raw materials, which would be more expensive in the high cost areas such as London, than in the lower cost areas of, say, Wales or Northern Ireland.

According to the data, the regional variations are approximately in accordance with Table 5.2 (taking Outer London costs as 100). As the anticipated project is in Nottingham (East Midlands), in accordance with the data in Table 5.2, the rate of

Table 5.1 Functional unit areas and unit costs

Function	*Indicative functional unit area*	*Indicative functional unit cost*
Administrative, commercial protective service facilities (Uniclass D3)		
Office – air conditioned – low density cellular	15 to 20 m^2/person	£21,000–£32,000/person
Office – air conditioned – high density open plan	10 to 15 m^2/person	£18,000–£30,000/person

Table 5.2 Regional variation of tender prices (UK)

Region	*Relative tender prices*
Inner London	106
Outer London	100
Southeast England (excl. London)	94
Southwest England	90
West Midlands	90
East Midlands	89
East Anglia	91
Northwest England	89
Yorkshire & Humberside	93
Northern England	94
Scotland	94
Wales	87
Northern Ireland	72

£26,500 should be reduced by 11 per cent giving a regionally adjusted figure of £23,585 per person.

For an office building holding 250 staff, the total usable floor area would be 3,750 m^2 (15 m^2 per person × 250 people) with a ball-park estimated construction cost of £5,896,250 (i.e. 250 × £23,585). Care must be taken, however, that these costs include everything that is necessary for the client to pay out. The notes which accompany the table in the source book state that the costs include preliminaries at 16 per cent (which may be considered high) and contractor's overheads and profit. They will not include external works (as these costs depend on the particular details of the site), fittings, furniture and equipment (FFE) and professional fees. An allowance for these costs must be included separately, if required.

Additionally, the above costs were based on the year 2008. Therefore, the costs must be updated to the anticipated date of tender for the project under consideration. If the new building is required for, say, September 2013; allowing a 12 month construction period and 4–6 weeks for the tender evaluation period, the tenders should be submitted at the end of the second quarter 2012 (2Q2012), but as this date is just on the cusp, it may be better to take the figure for third quarter 2012 (3Q2012).

Table 5.3 shows the tender prices given by *Spon's Architects' and Builders' Price Book* 2012.

The anticipated tender price may be adjusted from the 2008 annual average figure of 534 to the 3Q2012 figure of 465 (although this is a forecast figure as the anticipated

Table 5.3 Index of tender prices for UK construction (1976 = 100) (P = Provisional figure; F = Forecast figure)

Year	*First Quarter*	*Second Quarter*	*Third Quarter*	*Fourth Quarter*	*Annual Average*
2006	480	485	494	506	491
2007	515	520	528	538	525
2008	543	547	541	505	534
2009	500	485	460	455	475
2010	457	454	452	450	453
2011	446 (P)	447 (F)	450 (F)	453 (F)	449 (F)
2012	457 (F)	461 (F)	465 (F)	469 (F)	463 (F)
2013	474 (F)	477 (F)	483 (F)	486 (F)	480 (F)

tender date is still in the future). This would be applied to the regionally adjusted figure of £5,896,250 to calculate a new estimate for the building to be constructed in the East Midlands with an anticipated tender date of 3Q2012. Note that the 2012 index number is lower, meaning that the construction costs are expected to reduce.

The order of cost estimate is now £5,134,375 (i.e. £5,896,250 × 465/534), which would be advised to the client, probably rounded to £5,150,000 as the accuracy of the data is likely to be ±5–10 per cent as a minimum.

The client would then be able to assess whether or not they are prepared to accept this estimated overall cost when compared to their available budget and also whether a building of the required size would fit onto the site available (or if one is not yet available, the client and their land agents would know the size of plot required).

The figures in Table 5.3 make interesting reading and show how deep the 'credit crunch' recession has hit the construction industry in the two years from mid-2008. The tender index drops from a high of 547 in 2Q2008 to a low of 450 in 4Q2010, which represents a drop of 18 per cent in tender prices over two and a half years. Compare this with the index of building costs in Table 5.4 over the same period taken from the same publication.

During the same period (2Q2008 to 4Q2010), building costs (i.e. wages and materials costs to the contractor rose from an index of 742 to 798 (Table 5.4), although this last figure is provisional. This represents an increase in costs to contractors of

Table 5.4 Index of building costs for UK construction (1976 = 100) (P = Provisional figure; F = Forecast figure)

Year	*First Quarter*	*Second Quarter*	*Third Quarter*	*Fourth Quarter*	*Annual Average*
2006	664	670	694	699	682
2007	703	707	730	730	718
2008	733	742	781	780	759
2009	773	770	771	778	773
2010	780	793	797	798 (P)	792 (P)
2011	807 (P)	817 (F)	831 (F)	831 (F)	821 (F)
2012	837 (F)	844 (F)	858 (F)	869 (F)	852 (F)
2013	876 (F)	884 (F)	903 (F)	903 (F)	892 (F)

7.5 per cent at a time when tender prices reduced by 18 per cent. This clearly shows that the number of 'suicide bids' increased in this period as contractors reduced their prices in a desperate attempt to maintain a flow of work and it is therefore little surprise that the number of insolvencies in the construction industry rose sharply in this period, in all areas such as voluntary liquidations, receiverships and compulsory liquidations.

5.2 Design brief and initial formal cost plan

A design brief is a comprehensive document which should be produced jointly by the client and the lead designer and should be focused on the desired function that the building is intended to achieve. Design briefs are also important in order to ensure that the designer fully understands the functions that are required to be performed by the proposed building or facility and keeps these functions constantly in mind as the design progresses.

The brief will be produced at the beginning of the design stage, but is not a static document and will be continually reappraised as alternative design solutions are put forward and also due to any changes brought about by the client themselves – a process known as 'firming up the brief': this is intended to result in a clear set of instructions setting out the goals and objectives of the project as well as the detailed requirements. This may specify the number and size of space that is required, the relationship between the various types of space, the finishes, furniture and equipment required and the environmental conditions required, which will include temperature range, humidity, air movement, sound proofing etc. all of which will have an effect on the comfort conditions of the eventual occupants.

In complex building, such as hospitals, a design brief may be produced on a room-by-room basis and will also need to consider the building services requirements in some detail. A major rule of thumb in design is that 'form follows function', so an analysis of the function will be a pre-requisite for any effective design solution.

Therefore, for all construction projects, the client must brief the various participants about what is expected of the proposed building. Usually, there will be a formal or written brief or series of briefs that may form part of the tender documents, but in all cases, the client and the construction industry should explore, develop and communicate the client's requirements.

As mentioned above, the briefing should not be a static process but should take place throughout the construction process from project inception to completion. The client needs to be actively involved at all stages in order to ensure that the project meets their initial and changing requirements. Critical decisions are often taken during the early stages of the project which will affect the entire subsequent project and full client participation in these decisions is essential with adequate time and resources allocated to the process. A great deal depends on the interpersonal and managerial skills of the leaders of the briefing process and these must be developed to meet the demands of a particular project and set of participants. Factors such as client experience, organisational complexity and culture, rate of organisational change, project complexity and degree of project development all need to be taken into full consideration.

5.2.1 Key questions

Key questions that must be addressed during the design brief stage include:

- Has the client's objectives and/or business case been established in detail?
- Have other means of achieving them been evaluated before deciding to build?
- Has time been spent at the beginning of the process to define what is wanted, when and for how long?
- Is there an established budget and/or time constraint?
- Has there been a prioritisation between time, cost and quality?
- Has care been taken to choose the right people to represent and advise the client, with appropriate qualifications and experience and ability to work together?
- Have all risks been quantified and included within the budget?
- Have all costs-in-use (whole life costing) been identified?
- Have all construction options been identified?

5.2.2 Initial formal cost plan

The initial cost plan is slight misnomer, as no 'planning' is currently being carried out. A more correct description would be a *costing of the brief*, as all of the client's objectives, requirements and business case should have been established in the brief and a preliminary cost estimate is consequently developed. This cost estimate would use the project characteristics and compare them with the characteristics of similar projects which have been previously constructed, which is slightly different from the order of cost estimate, as this uses average costs per functional unit.

The initial formal cost plan is therefore based on the actual size of the building, in terms of total square metres of floor area. The estimate will be in elemental format and the costs based on historical actual costs which clearly need to be updated from when they were generated to the anticipated date of the project under consideration.

The list of building elements, taken from Uniclass (an industry standard system) is given in Table 5.5. Given the level of design of the building at this stage, it may not be possible to use all of the different elements, but the cost plan should be constructed in such a way that each section can be expanded as required, as the design progresses and details are firmed up.

What happens with ECI?

This is the stage at which a contractor would be appointed under an EPC or EPCM arrangement (see sections 3.3.1 and 3.3.2 for a definition of these arrangements). This means that the client has already carried out some preliminary schematic design (i.e. 'front-end' engineering and design – FEED) and will know the overall size, shape and major features of the project, including performance criteria, but has not yet carried out a detailed design. Early contractor involvement will allow the contractor to take over this schematic or front-end design and develop it into working drawings suitable for construction. The contractor also takes over the *responsibility* for the design, and must therefore ensure that they have the appropriate insurances in place as the legal liability of a designer is often greater than that of a constructor.

Whilst still being within the overall definition of design-build, contractors engaged after the design brief/FEED stage do not have the same opportunity for alternative

Table 5.5 Uniclass building elements and work sections

Section G – Building Elements	**Section J – Work Sections for Buildings**
G1 - Site preparation	JA - Preliminaries/general conditions
G11 - Site clearance	JB - Complete buildings/structures/units
G12 - Ground contouring	JC - Existing site/buildings/services
G13 - Stabilisation	JD - Groundwork
G2 - Fabric: complete elements	JE - In situ concrete/large precast concrete
G21 - Foundations	JE0 - Concrete construction generally
G22 - Floors	JE1 - Mixing/casting/curing/spraying in situ concrete
G23 - Stairs	JE2 - Formwork
G24 - Roofs	JE3 - Reinforcement
G25 - Walls	JE4 - In situ concrete sundries
G26 - Frame/isolated structural members	JE5 - Structural precast concrete
G3 - Fabric: parts of elements	JE6 - Composite construction
G31 - Carcass/structure/fabric	JF - Masonry
G32 - Openings	JF1 - Brick/block walling
G33 - Internal finishes	JF2 - Stone walling
G34 - Other parts of fabric elements	JF3 - Masonry accessories
G4 - Fittings, furniture and equipment (FFE)	JG - Structural/carcassing metal/timber
G41 - Circulation FFE	JG1 - Structural/carcassing metal
G42 - Rest, work FFE	JG10 - Structural steel framing
G43 - Culinary FFE	JG11 - Structural aluminium framing
G44 - Sanitary, hygiene FFE	JG12 - Isolated structural metal members
G45 - Cleaning, maintenance FFE	JG2 - Structural/carcassing timber
G46 - Storage, screening FFE	JG3 - Metal/timber decking
G47 - Works of art, soft furnishings	JH - Cladding/covering
G48 - Special activity FFE	JJ - Waterproofing
G49 - Other FFE	JK - Linings/sheathing/dry partitioning
G5 - Services: complete elements	JK1 - Rigid sheet sheathing/linings
G50 - Water supply	JK2 - Timber board/strip linings
G51 - Gas supply	JK3 - Dry partitions
G52 - Heating/ventilation/air conditioning (HVAC)	JK4 - False ceilings/floors
G53 - Electric power	JL - Windows/doors/stairs
G54 - Lighting	JM - Surface finishes
G55 - Communications	JN - Furniture/equipment
G56 - Transport	JP - Building fabric sundries
G57 - Protection	JQ - Paving/planting/fencing/site furniture
G58 - Removal/disposal	JR - Disposal systems
G59 - Other services elements	JS - Piped supply systems
G6 - Services: parts of elements	JT - Mechanical heating/cooling/ refrigeration systems
G61 - Energy generation/storage/conversion	JU - Ventilation/air conditioning systems
G62 - Non-energy treatment/storage	JV - Electrical supply/power/lighting systems
G63 - Distribution	JW - Communications/security/control systems
G64 - Terminals	JX - Transport systems
G65 - Package units	JY - Services reference specification
G66 - Monitoring and control	JZ - Building fabric reference specification
G69 - Other parts of services elements	
G7 - External/site works	
G71 - Surface treatment	
G72 - Enclosure/division	
G73 - Special purpose works	
G74 - Fittings/furniture/equipment	
G75 - Mains supply	
G76 - External distributed services	
G77 - Site/underground drainage	

design decisions, as a lot of the big decisions have already been taken and therefore the room for design cost efficiencies are reduced. The same point can also be made for cost savings through the buildability of the design, although cost savings through alternative specifications are still available.

Example

Continuing with the example of the Nottingham Office Block:

We now have a total estimate of £5,150,000 for the building, which has been based on functional unit costs as this was the best available information at the business justification stage of the design. Now that the actual building design is being developed, the preliminary cost estimate can be prepared with actual design information of the building under consideration. This is the beginning of the cost planning stage where the elemental format will be used, as shown in Figure 5.1. At present the design information is fairly sketchy, so the elemental costs are shown related to the gross internal floor area (GIFA) of the entire building. This requires that the building is of a fairly standard design – for example, the cost of the external walls will depend greatly on the floor to ceiling height, if the ceiling height is great, then the walls will be more expensive in relation to the floor area. However, showing the costs of the walls as a function of the floor area makes no allowance for variable floor to ceiling height.

As the cost planning stage progresses, the elemental costs will be shown in terms of their own quantities.

Figure 5.1 shows a standard calculation for the initial cost plan using costs per square metre of gross internal floor area. The overall costs of £5,167,345 is comfortably close to the order of cost estimate of £5,150,000, although there are several allowances made in the estimate, for FFE, contingencies and design fees, which will need careful monitoring during the cost planning stage.

The design stage can now progress and the cost estimate will now be progressively updated and firmed up in this elemental format.

5.3 Design development and updated formal cost plan

During this stage, the schematic design is progressed through to full working drawings. Clearly, at the end of the previous stage, the schematic design together with the initial formal cost plan should have been signed off/approved by the client or their authorised representatives (i.e. the project manager) before any further design development takes place.

The main components of the design development stage include:

a) development of the design details which have not been fully considered at the schematic design stage
b) confirming the spatial coordination of the design elements
c) preparation of the working drawings, construction details and specifications
d) confirmation of capital cost, which should ideally be carried out in stages as the design progresses (cost planning) to ensure that the estimated costs remain within the approved budget and any other projections of expenditure.

This activity therefore involves the ongoing development and refinement of the approved schematic design by all design consultants and the incorporation of any

NOTTINGHAM OFFICE BLOCK - PROJECT COST MODEL			
	ELEMENT	GIFA = 3750 m²	
		COST PER m² GIFA	ELEMENT COST
1	**SUBSTRUCTURE**	67.50	253,125
2	**SUPERSTRUCTURE**		
	Frame & Upper Floors	160.00	600,000
	Roof	47.00	176,250
	Stairs	16.00	60,000
	External Walls incl Windows & Doors	150.00	562,500
	Internal Walls and Partitions	28.00	105,000
	Internal Doors	45.00	168,750
	Total Superstructure		**1,672,500**
3	**FINISHES**		
	Wall Finishes	42.00	157,500
	Floor Finishes	43.00	161,250
	Ceiling Finishes	22.00	82,500
	Total Finishes		**401,250**
4	**FITTINGS AND FURNISHINGS**	allow	250,000
5	**SERVICES**		
	Sanitary Appliances	1.00	3,750
	Services Equipment	30.00	112,500
	Disposal Installations	5.00	18,750
	Water Installations	11.00	41,250
	Space Heating & Air Conditioning	90.00	337,500
	Electrical Installations	95.00	356,250
	Lift & Conveyor Installations	20.00	75,000
	Fire & Lightning Protection	3.00	11,250
	Communications & Security	28.00	105,000
	Builder's Work in Connection	3.00	11,250
	Commissioning	1.00	3,750
	Total Services		**1,076,250**
	BUILDING SUB-TOTAL		**3,653,125**
6	**EXTERNAL WORKS**		
	Site Works		
	Drainage		
	External Services		
	Minor Building Works		
	Demolition and Work Outside Site		
	Total External Works	15%	**547,969**
7	PRELIMINARIES	11%	462,120
8	CONTINGENCIES	5%	210,055
9	DESIGN FEES	7%	294,077
	TOTAL ESTIMATED PROJECT COSTS		**5,167,345**

Figure 5.1 Initial formal cost plan

further client requirements into the design. It would also entail the final design coordination and details of structural systems, MEP, HVAC and other building services together with the selection of materials and finishes and resolving any other design issues which remain outstanding from the schematic design stage.

On completion of the design development stage, the client will be provided with a completed design for the project, setting out all the necessary aspects to ensure the description of the planned works meets the client's delivery requirements, business needs and functional parameters. Design development should resolve all outstanding design issues to a stage which is ready for full accurate lump sum calculations by the tendering contractors.

As mentioned previously, design decisions taken at this stage *generate* the costs of a project, which are then *incurred* during the construction stage. Therefore, in the selection of materials, plant, equipment and finishes, consideration should be given to robustness, life-span and ease of operation. In the same way, design decisions can and do affect the buildability of a project; therefore it is significantly in the client's interests to ensure that a detailed knowledge of construction methods and programming are available during this design development stage, which is a point made by every design-build contractor at every opportunity.

Typical documentation required at the completion of this stage would include the following:

- site plan (1:500) including site development and any phased work
- plans, sections and elevations (1:100)
- room layouts (1:50) including necessary FFE
- construction sections (1:50) showing details at junctions of walls and floors, ceilings, stairs etc.
- preliminary construction details (1:20 and 1:10)
- site works and landscaping layouts (1:200) including stormwater, paving, car parking, excavation/fill, planting, engineering site services etc.
- MEP Services (1:100) including HVAC, lifts, plant rooms etc.
- FFE as required in the contract
- communications, security and data cabling
- building management, monitoring and control systems
- reconciliation of floor areas and other key design parameters against previous approvals
- certifications and other statutory building regulation approvals, depending on locality
- confirmation of any staged construction works and contractor's site establishment/ lay-down areas.

5.3.1 Key questions

Key questions that must be addressed during the design development stage include:

- Are the client's objectives and/or business case still being met by the detailed design completed at the end of this stage?
- Have other means of achieving them been analysed and evaluated during the stage (through a value engineering exercise)?

- Has time been spent throughout the process to define precisely what is wanted and by when (constant review process)?
- Is the established budget and time constraint still valid?
- Is the design being progressed methodically and in elemental stages?
- Is there constant and thorough collaboration between the different designers to ensure that there is full and proper coordination (e.g. architectural, mechanical and electrical drawings are designed, and checked, to fit together)?
- Have all residual risks been quantified and included within the final pre-tender estimate?
- Have all costs-in-use (whole life costing) been identified?
- Have all construction options been allowed for (if appropriate)?

5.3.2 Updated formal cost plan

In terms of estimating construction work and providing value for money, the design development stage is probably the most important stage of the whole project. As stated above, decisions made at this stage generate the costs in a project, and those decisions will have been made depending on whether speed, quality or cost is the major project criterion, or whether two are just as important, or they are all equally important. In other words, the project designers will position the project somewhere in the triangle shown in Figure 5.2.

In terms of the project estimate, the design development stage starts with the cost estimate produced at the end of the design brief stage, which should have been formally accepted by the client before progressing from the briefing stage. This estimate was called the initial formal cost plan in section 5.2.2. above.

Further cost estimates will be produced at key stages throughout the design development, as design decisions are made and incorporated into the drawings and specifications. It is vital that the cost estimates are updated on a regular basis, to

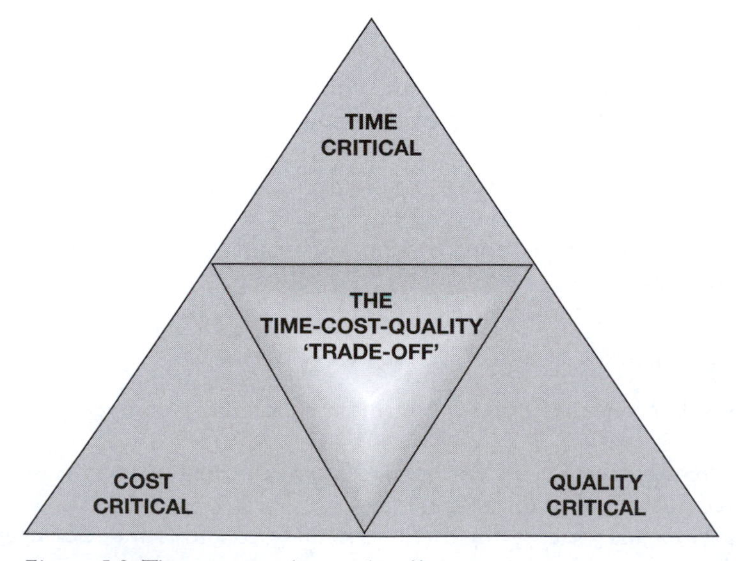

Figure 5.2 Time-cost-quality trade-off

ensure the design development does not get out of hand and the elemental budget parameters (initially set in the initial cost plan) are maintained and respected. Architects and engineers are fine fellows under normal circumstances, but they do not always keep their design decisions within cost allowances if they feel that a different option is more aesthetically pleasing or has better performance characteristics. There is nothing wrong with this per se, but if cost allowances are likely to be exceeded, the client should be given the opportunity to accept the higher cost as soon as possible, or savings could be found in other areas or elements of the project.

The design development stage is also the ideal stage for value engineering (VE), which is a systematic method to improve the 'value' of goods and services by relating their costs to their functional performance. Value, therefore, is the ratio of function to cost and can be increased by either improving the functional performance or reducing the cost. Obviously, the basic functions of a building should be preserved and preferably enhanced during this process; therefore VE is not just a cost-cutting exercise. Value engineering also applies to life cycle cost (LCC) analyses by assessing the cost-in-use of the products, materials or systems being considered.

A design development report will be prepared at regular intervals, which is intended to demonstrate that the issues of planning, design, materials selection, construction, staging and phasing, services integration and coordination etc. have been addressed and integrated into the proposal to ensure an effective project outcome.

The updated formal cost plan will normally be in the same elemental format as the initial formal cost plan and its purpose is to refine the cost estimate to a greater level of accuracy by taking account of the more detailed design information which becomes available as the design progresses. At the end of the detailed design stage, the cost estimate should be within ±10 per cent of the actual costs, although this can vary depending on the procurement route chosen and the market competitiveness at tender stage.

Figure 5.3 gives a graphical illustration of how the cost estimate is refined throughout the design stage.

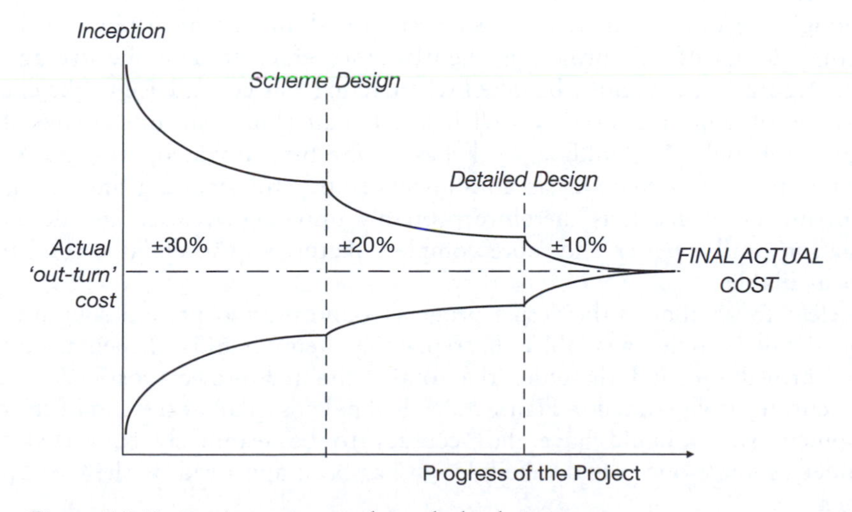

Figure 5.3 Estimating accuracy through the design stage

What happens with ECI?

This stage is where the design-build arrangement can really come into its own. If the company who will actually perform the construction activity is involved in the design development, they will want to ensure that all coordination issues and buildability issues are dealt with at this stage. This is especially true if they are working to a lump sum price, so any inefficiencies leading to extra costs will have a direct impact on their profit margins.

Traditionally, the design of the architecture, structure, building services (MEP) and internal transportation (elevators, escalators etc.) are carried out by separate organisations and often at the same time. Therefore, there is clearly an opportunity for coordination problems between the different elements of the building and it has only been relatively recently that the role of design manager has been created to try to solve these issues. A design-build contractor takes on the role of design manager to ensure this coordination. Additionally, the construction organisation employs specialist skills in planning, programming, HSE and other buildability issues, all of which can have the effect of increasing the efficiency of the construction process, thereby reducing costs and improving safe working methods. It was a deliberate policy in the UK health and safety industry to ensure that the H&S regulations included the design stage – Construction (Design and Management) Regulations (CDM).

Example

Continuing with the example of the Nottingham Office Block:

As the design progresses, the initial formal cost plan shown in Figure 5.1 can be updated through an approximate measurement exercise of the design, usually carried out by the project quantity surveyors/cost consultants. For a design-build contract or where the contractor has been appointed early, this procedure can be carried out by the contractor's estimating department.

As more project design detail becomes known, the cost plan will therefore be updated as shown in Figure 5.4.

Although the total estimated project costs are shown to have reduced, because the actual design of this project is slightly more efficient than the average costs given in Figure 5.1, care must be taken when giving this cost advice to the client, as other areas of project costs may well increase, cancelling out any savings shown. In Figure 5.4, only the building work has so far been updated, with the services and external works remaining as an allowance and still awaiting final design and specification decisions. It is therefore probably unwise to release any design cost information until a better and more complete picture of the updated total project costs is available.

It is clear to see that as the design progresses and more definite design and specification detail becomes available, the remaining items in this elemental cost plan can be 'firmed up' and therefore the total estimated project costs will become more accurate, as illustrated in Figure 5.3. The final cost plan at the end of the design development stage should have the accuracy to be seamlessly converted to the pre-tender estimate once the detailed design has been approved by the client/project manager.

NOTTINGHAM OFFICE BLOCK - ELEMENTAL COST PLAN						
		GIFA			3750 m²	
ELEMENT		ELEMENT UNIT QUANTITY		ELEMENT UNIT RATE	TOTAL ELEMENT COST	COST PER m² GIFA
1	**SUBSTRUCTURE**	2,100		130.00	273,000	72.80
2	SUPERSTRUCTURE					
2A	R.C. Frame & Upper Floors	4,000	m²	140.00	560,000	
2B	Upper floors	incl				
2C	Roof	800	m²	75.00	60,000	
2D	Stairs	10	no.	5,500.00	55,000	
2E	External Walls	960	m²	70.00	67,200	
2F	External Windows and Doors	750	m²	350.00	262,500	
2G	Internal Walls and Partitions	2,500	m²	50.00	125,000	
2H	Internal Doors	500	no.	350.00	175,000	
	Total Superstructure				**1,304,700**	347.92
3	FINISHES					
3A	Wall Finishes	5,000	m²	35.00	175,000	
3B	Floor Finishes	4,000	m²	40.00	160,000	
3C	Ceiling Finishes	5,000	m²	23.00	115,000	
	Total Finishes				**450,000**	120.00
4	**FITTINGS AND FURNISHINGS**	allow			**250,000**	66.67
5	SERVICES					
	Sanitary Appliances		item		3,750	
	Services Equipment		item		112,500	
	Disposal Installations		item		18,750	
	Water Installations		item		41,250	
	Space Heating & Air Conditioning		item		337,500	
	Electrical Installations		item		356,250	
	Lift & Conveyor Installations		item		75,000	
	Fire & Lightning Protection		item		11,250	
	Communications & Security		item		105,000	
	Builder's Work in Connection		item		11,250	
	Commissioning		item		3,750	
	Total Services				**1,076,250**	287.00
	BUILDING SUB-TOTAL				**3,353,950**	894.39
6	EXTERNAL WORKS					
6A	Site Works					
6B	Drainage					
6C	External Services					
6D	Minor Building Works					
6E	Demolition and Work Outside Site					
	Total External Works		15%		503,093	134.16
7	PRELIMINARIES		11%		424,275	
8	CONTINGENCIES		5%		192,852	
9	DESIGN FEES				295,000	
	TOTAL CONTRACT SUM				**4,769,169**	1,271.78

Figure 5.4 Elemental cost plan proforma

5.4 Approval of detailed design and formal cost plan

This is not so much a stage in the design process as a decision point, i.e. the third traffic light in Figure 4.1 in Chapter 4. As stated in the RIBA Plan of Work in Chapter 4, the final design is signed off by the client, which allows the tender documents to be prepared thus allowing the tendering contractors to accurately price the works. The typical design documents required at this stage would include the list in section 5.3 above, which should be both accurate and complete, as well as, hopefully, having been coordinated and integrated together. All drawings should be numbered in accordance with an industry standard system, such as Uniclass or CI-SfB. The specification document(s) should also be referenced using the same system to allow easy cross-referencing between all design information. This integrated system of cross-referencing was introduced into the UK industry in an initiative called 'Co-ordinated Project Information' in the late 1980s and is now part of the Uniclass system. Integrated cross referencing is essential on any project using Building Information Modelling (BIM).

5.4.1 Key questions

Key questions that must be addressed at the completion of the detailed design stage include:

- Are the client's objectives and/or business case still being met by this detailed design?
- Has the design been properly and thoroughly coordinated between the different designers (e.g. architectural, structural, mechanical and electrical drawings fit together and work in practice)?
- Have all residual risks been quantified and included within the final pre-tender estimate?
- Have all costs-in-use (whole life costing) been identified and reported to the client?
- Is the established budget and time constraint still valid?
- Has the budget been allocated effectively between the various elements of the building (in the cost plan)?
- Do the elemental cost estimates reflect good value rather than cheap price?
- Has the elemental cost plan been 'cost checked' by an approximate measurement exercise of the finished drawings?

5.4.2 Final cost plan

The final cost plan will normally be in the same elemental format as the previous cost plans and its purpose is to complete the cost estimating/cost planning process by calculating the final pre-tender estimate (often referred to as PTE) to the best level of accuracy by taking account of all the detailed design information included in the drawings and specifications. As stated in section 5.3.2, this cost estimate should be within ±10 per cent of the actual costs, although many clients will require a more accurate estimate of ±5 per cent with the estimate preferably being on the high side of the actual tender price submitted by the lowest tenderer. The reason for this is that the client will be allocating a budget to this project from capital funds which have alternative uses (and opportunity costs). Therefore, they need to know in advance how much capital expenditure should be allocated. However, in a very competitive and uncertain market, it is notoriously difficult to assess tender prices, especially when the tender may not be required until some time in the future – as can be seen from section 5.1 and Table 5.3.

The final cost plan represents the last stage of estimating using the elemental method. As mentioned above, the final elemental cost plan will be 'cost checked' using unit rate measurement techniques (i.e. the items that would appear in the bills of quantities – BOQs). From this point in the project, all estimates, tenders and pricing information will use the items as stated in the BOQ or other pricing document – mostly as items measured in accordance with one of the various standard methods of measurement (SMMs), but may also be in activity schedule format if that is how the project pricing document is formatted. However, for approximate estimates, these items are often aggregated together for simplicity, to create what are commonly known as 'builder's quantities'.

On completion of this construction estimate, a representative construction schedule can then be developed to create a total project cost summary (TPCS) for the client's internal purposes. The schedule and its logic can be used as a quality check of the estimate in relation to duration, labour requirements etc. The schedule logic may cause the estimate to be revised based upon the specific needs of the project and the schedule should be developed in sufficient detail to display the critical and near critical construction elements as well as critical concurrent activities.

The TPCS calculated at this stage will be used for project authorisation as part of the client's financial procedures and also forms the basis for allowable cost increases without the need for continual re-authorisation. The total project cost (TPC) at the time the project is authorised by the client becomes the baseline cost estimate (BCE) and represents the scope and schedule established at this stage. It is therefore clearly vital that the estimate established is both accurate and sufficient, as most corporate clients will allocate capital expenditure at the beginning of the year and do not wish to alter those budgets unnecessarily.

Allied to the BCE is another concept, known as the baseline project schedule (BPS). Information available during the design stage should allow a preliminary project schedule to be established as well as the preliminary costs. Clearly, this will be easier with early contractor involvement, as the contractor brings some very well-developed project planning skills to the party. The BPS is the initial project schedule to be updated as the detail is developed both in the design stage and to a lesser extent in the construction stage.

What happens with ECI?

When a contractor has been appointed before the completion of the detailed design, the tender stage is consequently shifted to earlier in the project. Therefore, the final cost plan serves a slightly different purpose in that it will form the basis of the contract price with the contractor.

Traditionally, the contractor will submit a tender price calculated from design information and tender documents produced after the design has been completed. See Chapter 6 for a fuller discussion of this process. The client will evaluate these tenders, accept one of them (after some post-tender negotiations) and a valid contract will be formed and both parties will know the scope of work, cost and time obligations. However, when a contractor has been appointed early, the contract is formed before the scope of works, time and cost obligations are fully detailed and agreed. Therefore ECI is often carried out in a two-stage process:

- Stage 1 – The contractor is appointed on an outline scope (design brief) with a cost estimate (initial cost plan) and outline programme.
- Stage 2 – The terms of the contract are 'firmed up' at design completion stage when an accurate scope of works and detailed cost estimate or lump sum can be calculated.

It is clearly not the purpose of this book to discuss the contractual procedures of two-stage tendering and the reader is referred to separate books on building procurement. However, the cost estimate produced at design completion stage will often serve as the contract price in ECI contracts and there is generally no need to produce detailed bills of quantities since there is no further tendering procedure at this stage. The purposes that BOQs serve during the construction stage will need to be carried out in a slightly different way.

Example

The final cost plan is also sometimes termed the pre-tender estimate, as it is the final estimate at the pre-tender stage (assuming of course that the contractor has not yet been appointed). When this estimate has been completed, it will be presented to the client for final approval and the go-ahead for release of the tender documents to the contractors. This is the third traffic light in Figure 4.1 in Chapter 4.

Figure 5.5 shows the format of a final cost plan/pre-tender estimate for a multi-functional commercial and residential development including retail, offices and residential construction, with each type of construction being separately reported. This allows much easier analysis of the efficiency of the design as well as the possibility of splitting the project into separate bid packs for tendering.

5.5 Summary and tutorial questions

5.5.1 Summary

A recurrent theme in this book is the standard phases or stages that all projects must go through. For a project to be considered as being managed effectively these stages must be recognised and the techniques appropriate to each stage used by the project managers.

The inception and design stages are crucial to the project, as this is where the project costs are generated – from decisions made by the client, designers and specifiers who together will choose the quantity and quality of the building being proposed. The main sections of this stage have been well documented in Chapter 4, and in this chapter the most appropriate and effective estimating techniques are outlined in order to ensure that the best quality financial advice is given to the client depending on the quality and extent of information available.

At the business justification stage, the order of cost estimate will be calculated based on average costs per functional unit – example given in Table 5.1. Information is available to adjust these costs depending on the location of a potential project and the expected date of tender.

Name of client
Name of project
Date of estimate

CSI Ref.	Element	Parking			Retail			Mall - Circulation			Offices			Residential			External work			TOTAL		
		Total Cost	Cost/ m^2	%	Total Cost	Cost/ m^2	%	Total Cost	Cost/ m^2	%	Total Cost	Cost/ m^2	%	Total Cost	Cost/ m^2	%	Total Cost	Cost/ m^2	%	Total Cost	Cost/ m^2	%
A10	Foundations / Substructures																					
A20	Basement Construction																					
B10	Superstructure																					
B20	Exterior Closure / Façade																					
B30	Roofing																					
C10	Interior Construction																					
C20	Stairways																					
C30	Interior Finishes																					
D10	Conveying Systems																					
D20	Plumbing Systems																					
D30	HVAC																					
D40	Fire Protection Systems																					
D50	Electrical Systems																					
E10	Equipment																					
E20	Furnishings																					
F10	Special Construction																					
G10	Site Preparation																					
G20	Site Improvements / Landscaping																					
	Sub-Total																					
Z10	Prelims / General Req. 12%																					
	Building Cost @ 1Q 2009 US$																					
10.00%	Contingency (Parking Areas)																					
	Contingency (Remainder)																					
	Escalation @ 0.15% per month US$																					
	SUB TOTAL US$																					
	Built-up Areas																					

Exclusions :
Utilities connection & government fees
Demolition work
Escalation cost beyond (Date)
Design fees
LEED (Leadership in Energy and Environmental Design) & ESTIDAMA requirements
Commercial Pantry equipments

Note
Escalation is based on 6 months for design completion & 18 months mid construction period

Figure 5.5 Final cost plan/pre-tender estimate proforma

During the design brief stage, an initial formal cost plan will be produced based on element costs given as a cost per square metre of gross floor area, as the actual element quantities are not yet known. Figure 5.1 gives an example of this calculation in elemental format.

During the design development stage, an updated formal cost plan is produced which can now include the elemental unit quantities as these become available from the designers. This clearly is a working document which is continuously updated as the design progresses and will be much more accurate than the previous techniques as more detailed project information is being used. Figure 5.4 shows the format of the elemental cost plan.

At the end of the design development stage, when the actual design has been approved by the client or project manager, the final cost plan will be produced in elemental format. Because the design is now complete, this cost plan can be checked by measurement of the drawings to produce a pre-tender estimate, which will normally be in the format of the pricing document of the project being considered – either a bill of quantities or activity schedule.

The purpose of this whole process is to ensure that the client's original budget is both feasible and achievable.

5.5.2 Tutorial questions

1. What do you understand by the term 'functional unit costs'?
2. Explain the process of converting the functional unit costs to the anticipated cost of an actual project.
3. Discuss the various effects of reducing tender prices (Table 5.3) with increasing building costs (Table 5.4).
4. What are the assumptions which must be taken into consideration when using elemental unit costs as costs per square metre of gross floor area?
5. Outline the sources of cost data when deciding on the element unit rate.
6. What are the advantages and disadvantages of using measurement techniques for cost estimating at this stage in a project?
7. Discuss the main purposes of a pre-tender estimate.

6 Estimating techniques at design completion stage

6.1 Design information and tender documents

The design completion stage *should* be exactly that. The scope of works/detailed design *should* be complete, finished, fit for purpose and, most importantly for the contractor's estimating department, be sufficient to be able to calculate a lump sum fixed price with a high degree of certainty, which is what the contractor will be held to for the duration of the construction period. Unfortunately, and for a variety of reasons, this is an aspiration rather than a fact in the majority of construction projects.

The above paragraph assumes that the contractor has not been involved during the design stage, and in partial mitigation for the design team, it is notoriously very difficult to accurately predict how long a design is likely to take to achieve completion, whatever 'completion' means. Designers, like contractors, nowadays are often required to work on a fixed fee basis and therefore need to control the number of hours spent on a project design, in order to ensure some degree of profitability for their company. Under normal circumstances, the first 60 per cent of a design can be completed reasonably quickly, but as the design progresses through the technical design and production information stages, the numbers of individual design decisions increase exponentially (known as a 'decision tree'), therefore requiring greater resource input from the designers and consequently, more time to achieve completion. This reflects a standard Pareto curve or 80/20 rule, which can be graphically illustrated as in Figure 6.1.

So, in the majority of cases, the design will not be fully completed or developed by the time the design drawings need to be sent to the tendering contractors as part of the tender documents. This state of affairs will be acceptable if the tendering contractors know about it, are allowed sufficient time themselves to either complete the design or be able to assess the risks that they are expected to take on. Dream on. Again, in the majority of projects, the contractor will be given an incomplete design, tender documents which may be incorrect, ambiguous or contradictory and still be expected to submit a lump sum fixed price tender in a very short period of time. To add insult to injury, many clients will then engage in a process of 'reverse-auction', where the tenderers are shown the values of all the other tenders (although with names deleted) and are asked to reduce their prices, so that the client obtains the cheapest possible (fixed) price for their incomplete design. Somewhat surprisingly, clients and their design consultants are still shocked and disappointed when claims and disputes arise during the construction stage.

The above paragraph may be viewed as somewhat cynical, but many experienced contractors will agree that this warts-and-all description is what actually happens in

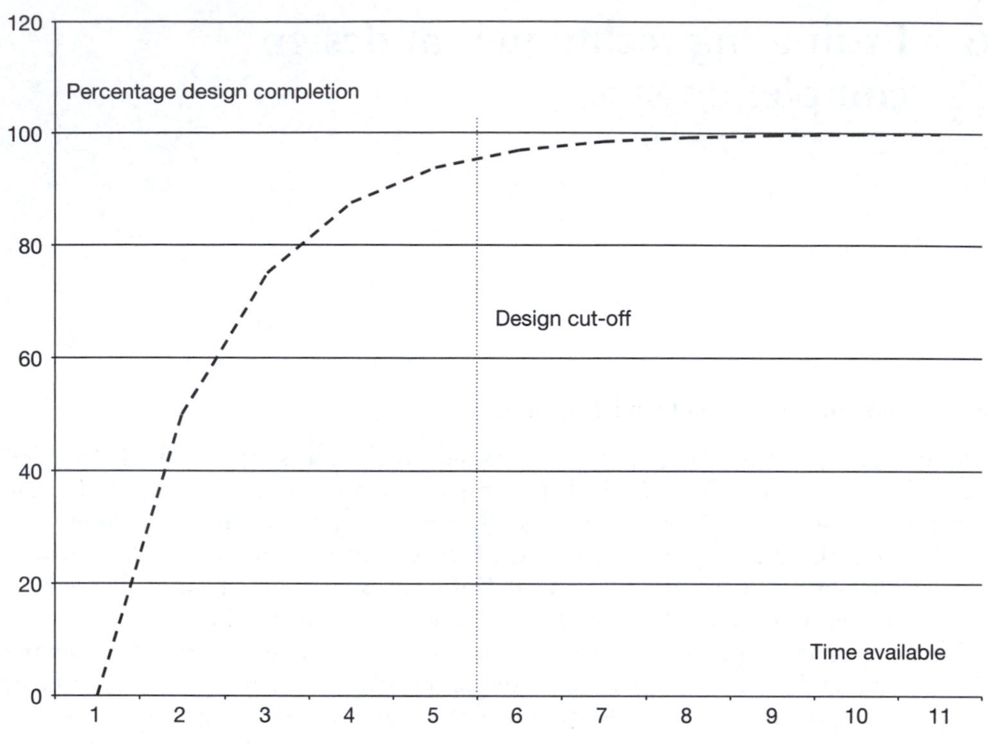

Figure 6.1 Pareto curve of design completion against time

practice, and it is therefore in the interests of all parties for the processes in each stage to be carried out as thoroughly and professionally as possible. Given the maxim that time spent in preparation is never wasted, a well-thought-through client's brief, a thorough and complete design, together with tender documents which accurately reflect the design, required specification and the approved procurement route will lead to a competitively priced tender and a reduced number of disputes in the construction stage of the project leading to increased project satisfaction for all concerned.

In terms of the tender documents received by the tendering contractors, this will depend on the procurement route which has been chosen and at what stage the contractor is to be involved in the project. In all cases, the tender documents should include:

a) all drawings showing the scope of works of the proposed contract
b) all documents showing the required quality of materials and workmanship (specifications)
c) a pricing document for the contractor to submit the bid (bills of quantities, bills of approximate quantities, schedule of rates or activity schedule)
d) the form of contract anticipated to be used
e) any specific tender instructions, such as:
   i) date of site visit/clarification meeting
   ii) date of submission of tender
   iii) person to whom all correspondence should be addressed
f) any other client-specific documents.

When the tenderers have received this information, they are in a position to start the process of building up the estimate, i.e. the anticipated actual costs of carrying out the work. There is a great deal of cost information in the public domain (such as *Spon's Architects' and Builders' Price Books*), which purport to give costs for items which have been measured in accordance with the standard methods of measurement (SMM/CESMM) and these costs may also be split into the component parts of labour, plant and materials. It is therefore very easy, on receipt of a full bill of quantities which has been produced in accordance with an SMM, to use these costs in an actual tender. However, there are several difficulties with this approach:

a) The pricing information is in the public domain, so everyone else has access to the information and clients have been known to receive two or more identical tenders.
b) Pricing books are notoriously out of date. The 2012 edition would have been produced and printed in mid-2011, therefore in a competitive, volatile market, the pricing information is indicative at best.
c) The prices are not the tenderer's own costs as they have not been built up from first principles, i.e. the tenderer's costs of *their* labour, *their* material costs and *their* plant costs. As stated in an earlier chapter, this is the main competitive differential between the tendering companies.

For the tendering contractors to build up the estimated costs from first principles, they will need to go through the following procedure:

- As each contractor will obviously work in a slightly different way, each tenderer will prepare a tender method statement showing how the work will be carried out, using the standard working practices of the company together with the plant and equipment available for the project, either on a hire or own basis, depending on the company policy. See section 6.2 below.
- Develop a tender programme or preliminary baseline schedule, to ensure that the project can be constructed during the contract period given by the client (i.e. the time period between the date for possession of the site and the required date for completion, which will have been given in the tender documents, including any required sectional completions). See section 6.3 below.
- The temporary works that will be required in order to construct the project. Drawings, specifications and bills of quantities normally only reflect the permanent or finished works. Any temporary works required to be able to construct the finished works are usually the responsibility of the contractor. See section 6.4 below.
- Plant requirements are again dependent on the contractor's company policy and the choice of individual items of plant and equipment will clearly affect the efficiency of operations and hence the cost. See section 6.5 below.

All of the above decisions are normally carried out concurrently during the earlier stages of the tender period, so that the costs can be brought together and converted into the unit rates and inserted into the bills of quantities. See section 6.6 below.

Of course, a bill of quantities is not the only way that a price can be built up for tendering purposes, and given the point in the previous paragraph, the BOQ is not a particularly good way of pricing construction work. The reason for this is that the tendering contractors have estimated their costs from first principles – i.e.:

a) How are we going to do the work? (method statement)
b) When is each operation going to be scheduled? (tender programme)

Therefore, their costs have been developed in terms of costs of operations or activities. However, they are now required to convert those costs into unit costs (i.e. cost/m^2 or cost/m^3 etc.), mainly because that's how it has always been done and it makes the consultant's life a bit easier in the tender evaluation stage and the construction stage, when approving interim valuations. However, many of the newer standard forms of contract, notably NEC3 Options A and C, allow an activity schedule to be the project pricing document, which means that the contractor has no need to convert the activity costs into unit rates, and is paid on completion of each completed activity during the construction stage. Unfortunately, things that sound too good to be true invariably are, and there are other difficulties associated with this approach which are unfortunately outside the remit of this book to discuss.

The remaining cost items for the contractor are mainly associated with the work of subcontractors, which is discussed in detail in section 6.7 below.

6.2 The contractor's method statement

A method statement is exactly what it says on the lid, i.e. a document that informs the relevant parties of how a task is to be performed, and of any requirements that may be an issue to enable the works to be carried out both correctly and safely.

Method statements are not new and they have been used for many years by contractors wishing to properly plan the works in order to know the time and cost implications of the works. In practice, during the construction works, events on a site may overtake the method statement procedure and the programme may therefore be affected because certain works are not completed at the scheduled time. It is therefore essential that method statements are compiled after taking on board site-specific requirements and are issued to all the relevant parties within an agreed timetable.

Construction method statements would include the following topics:

1 scope of works required in the project
2 site-specific details which may affect methods of construction, e.g. topography
3 method of working decided by the contractor
4 assessment of any significant risks
5 access points to the site (including potential unauthorised access points)
6 requirements for working at heights
7 use of substances hazardous to health
8 any required limitations on noise
9 any requirements for manual handling of goods and materials
10 any requirements for working in a hazardous or confined area
11 resources available to the contractor
12 control measures required by the client/consultant or local authority
13 requirements for personal protective equipment (PPE), especially special equipment
14 emergency arrangements

15 risk assessments of operations
16 who the information should be issued to
17 how the method statement will be monitored.

The importance of a properly prepared method statement cannot be overestimated and is crucial to the tender evaluation procedure and also to the health and safety of construction operations. When compiling a method statement, the contractor should take into account all facts regarding the site, which will often require a visit to the proposed site in order to ensure the method statement is as specific to the site conditions as practically possible.

The site manager, supervisor and workforce must have read and understood the method statement, which should have been issued prior to the work being carried out, and should also be given an opportunity to revise the method statement given their knowledge of the task to be performed. The method statement is to be included in the project health and safety file and this would be referred to, in the event of an incident occurring that requires an investigation. In this instance, should a method statement not be present, the employer and employee could be held accountable by any investigating body and may incur criminal proceedings under CDM and health and safety legislation in the UK.

Therefore, in summary, a method statement is designed to:

- identify the scope of works
- identify potential hazards
- allow the employee to understand the scope of work and the specific requirements
- inform the client of the specific requirements of the work
- form part of the project health and safety file.

6.2.1 Types of method statements

There are basically two types of method statement.

Generic method statement

A generic method statement is the most popular type, as they have already been structured into the most common sections. The generic method statement will have been developed from a template with the only changes being to site- or project-specific issues. It will list the basic activities for the operation under consideration but may not cover all eventualities or requirements for the operation. These types of method statement will usually only be acceptable for small work activities that involve very little risk to the employee or anyone affected by the work. An example of a generic method statement for a simple operation is shown in Table 6.1.

This generic method statement shows the basic tasks to be carried out but does not identify any risks associated with the work, although the required materials, plant and equipment are stated.

Project-specific method statement

A project-specific method statement will be developed, usually after visiting the site location to view and inspect where the operations and activities are to take place. This

Table 6.1 Generic method statement

METHOD STATEMENT FOR CONCRETING A LINE OF FENCE POSTS	
1	Excavate pit by hand for fence post to dimensions 450 × 450 × 300 mm.
2	Remove excavated material by dumper truck to temporary spoil heap on site – approx 200 m.
3	Lightly compact bottom of excavation.
4	Mix concrete in small petrol-driven motorised mixer adjacent to pit.
5	Place fence post in position ensuring it is vertical and aligned.
6	Pour concrete into excavated pit to 50 mm from top, ensuring the fence post is not moved.
7	Lightly tamp the concrete to ensure it has settled and the surface is flat and horizontal.
8	Support the fence post in position until the concrete has set.
9	When concrete has set, remove supports.
10	Check alignment of fence post.
11	Repeat with next post.

type of method statement clearly involves more effort in its production and consequently contains much more detail of the activity, the required resources and all associated risks.

The project-specific type method statement should identify all aspects of the work and inform the relevant parties of any requirements or hazards associated with the operations or activities. These need to be developed by competent people who are familiar with compiling method statements and have a detailed knowledge of the tasks to be performed. They should ideally have some experience in risk and hazard identification, as this topic will be required to be included within the project-specific document.

This type of method statement will show the task being carried out as per the generic method statement but will also identify other key issues, as shown in Table 6.2.

6.2.2 Advantages and disadvantages of method statements

As with most production operations, the key to effective work control is the definition of a scope of work within a work breakdown structure (WBS). If there is no accurate work scope, both the employer and employee would have difficulty in controlling the work adequately, and this is probably one example where the somewhat glib management-speak expression 'If it can't be measured, it can't be managed' has some merit. Defining the work to be performed impacts on all aspects of work control, it is not possible to set reliable timescales and resource requirements if the scope of work has not been defined adequately.

Advantages of construction method statements

- The operative has a greater knowledge of the project requirements prior to attending the workplace.
- The operative and the employer will have details of risks associated with the work.
- The operative can carry out the work knowing that the employer has taken the time to review the workplace and has taken measures to ensure the work can take place in a safe and efficient manner.
- The employer will have full confidence that the operative has all the relevant details to allow the work to proceed safely and efficiently.

Table 6.2 Project-specific method statement

METHOD STATEMENT FOR CONCRETING A LINE OF FENCE POSTS	
1	**Scope of works**
1.1	To erect fence posts as per Drawing XYZ in accordance with programme schedule.
2	**Site-specific details.**
2.1	The fence location is shown on Drawing XYZ at the site perimeter, adjacent to an existing residential housing estate. Excavation to be carried out by hand therefore no large items of plant are required.
2.2	Dumper truck available as required to transport spoil to temporary heap on site.
2.3	Concrete mixed by mobile petrol-driven concrete mixer moved to required location as necessary. Water available by hose from site.
2.4	Fence posts carried by hand to location – two persons.
3	**Method of working**
3.1	The method used to erect the fence posts will be as follows: • Inform all relevant parties of intention to carry out work. • Move small petrol-driven concrete mixer adjacent to work site location. • Ensure all small hand tools are available and materials for concrete. • Bring sufficient fence posts from stores for one morning/afternoon work. • Select location of post foundation. • Ensure all operatives have required PPE. • Excavate pit by hand 450 × 450 × 300 mm. Deposit excavated material adjacent to pit. • Roughly compact bottom and sides of excavation. • Whilst excavation is being carried out, mix concrete to required specification in small petrol-driven mixer. • Place fence post in position and pour concrete by hand to 50 mm from top. • Check accuracy of post – position and vertical alignment. • Support the fence posts in position with temporary shoring. • The site foreman will be requested to view the completed work.
3.2	Upon completing the work as shown on the daily worksheet all posts will be checked for accuracy before concrete is finally set. Any unused posts will be returned to secure site stores. Concrete mixer will be cleaned and washed and returned to secure site storage.
4	**Assessment of significant risks** • Access requirements – access will be required to the secure stores (authorised personnel only). • Working at heights – not applicable. • Control of substances hazardous to health – cement powder should only be handled wearing protective gloves as required in PPE instructions below. • Noise – the noise generated by the concrete mixer may require ear protection. To be available in site stores. • Manual handling – cement powder, coarse aggregate and fine aggregate will be required to be manually placed in the concrete mixer. The mixer itself will be required to be manually moved to different positions as work proceeds. • Working in hazardous areas – not applicable.

(Continued)

Table 6.2 (Continued)

METHOD STATEMENT FOR CONCRETING A LINE OF FENCE POSTS	
5	**Resources** The gang shall consist of one craftsman and two labourers.
6	**Control measures** A permit to work shall be obtained from the site manager.
7	**Personal protective equipment (PPE)** The following PPE shall be issued and worn by the operatives during the works: • rubber soled steel toecap footwear • full length overalls • waterproof high visibility jacket • waterproof gloves • hard hat • eye protection as required.
8	**Emergency arrangements** Emergency arrangements are detailed on the permit to work and are as per site requirements and company health and safety document, which is available for viewing in the site offices.
9	**Risk assessment** The risk involved whilst carrying out the work has been deemed as significantly low; therefore, no special measures are required. Reviewed as work proceeds.
10	**Information issued to** The method statement will be issued for approval to the site manager, the workshop manager and the operative; a copy shall be kept in the site welfare facilities on view, on the 'work ongoing board'.
11	**Monitoring and compliance**
11.1	The ganger shall complete a daily sign-off sheet and issue it to the site manager; the forecourt manager shall inspect all areas used by the operatives to ensure that all works are progressing as per the permit to work and this method statement, as required.
11.2	The site manager shall issue corrective action directives in the first instance and take appropriate action should the need arise.

- The client has a greater knowledge of what the contractor/operative will be doing and what activities, if any, they have to provide assistance for.
- The issue of an approved project-specific method statement will allow the employer and contractor to understand the division of responsibilities between them.
- The operatives have to adhere to the method statement and cannot diverge from the agreed method of working unless another method statement is issued and approved; this cuts out contractors being able to cut corners to save time and money.
- Proper and effective work controls will reduce any abortive works and will improve teamwork.
- The issuing of method statements will improve operative and company efficiency, which will have a direct effect on profitability.
- The requirements of the CDM Regulations will have been met.

Disadvantages of construction method statements

- The company would have to train people to become competent at compiling the statements.
- The usual late order for the works and time restraints associated due to the late placement of the order will put pressure on the contractor to try to use a generic method statement.
- The method statement will be commented on, and may require revising in line with the comments and would require to be reissued.
- The company would have to direct labour to the duty of compiling the documents, remember if they are to be site specific, a visit to the workplace will usually be required and this may not have been included in the budget at the tender stage.
- The operatives have to adhere to the method statement and cannot diverge from the agreed method of working unless another method statement is issued and approved, this cuts out contractors being able to cut corners to save time and money.

6.3 The contractor's pre-tender programme/baseline schedule

A pre-tender programme is exactly that. It is an estimate of when the work will be carried out starting at the commencement date given in the tender documents and ending on or before the completion date given in the tender documents. Unfortunately, contractors are rarely at liberty to proceed with the works at their own pace (if they are fortunate enough to find themselves in this position, it is known as 'time at large'). Therefore, the construction must be fitted in the time slot between the commencement date and completion date given to them. This time slot is commonly known as the 'contract period' and the duration of the actual construction works is commonly known as the 'construction period'. The two should not be confused and an analysis of the two periods will form the basis of the contractor's applications for extensions of time for delay and/or disruption.

If the contractor chooses to complete the works before the contractual date for completion, the time difference between the actual completion date and the required completion date is termed the 'project float' and many contractual arguments have occurred related to this concept. However, this is outside the scope of this book, suffice it to say that at tender stage, the contractor and its estimating department will try to ensure that the works are scheduled to be completed on or before the required date, by often incorporating a dummy activity such as 'tidying up the site' to ensure there is no project float shown in the programme.

The overall programme schedule, once it has been decided, needs to be separated down into the main operational stages of the project, such as the list of work sections given in Table 5.5 in Chapter 5. Information about how long each of these periods will take is essential to the estimator, which will enable a calculation of:

1 labour requirements
2 site accommodation
3 mechanical plant and equipment
4 temporary works and falsework
5 work affected by seasonal weather changes.

Specifically related to the *direct costs* of the works (i.e. those labour, plant and material costs of the actual works to be built), the estimator will assess if they can be built under normal productivity rates, or if additional resources are required. Once this decision is made, the labour requirements are passed to the project planner, who will incorporate them into the tender programme, in order to develop a histogram of requirements at each point in time in the proposed programme. If there is a bottleneck or the proposed histogram varies wildly, then the project planner may carry out a resource levelling exercise, if there is sufficient leeway (or float) in the various individual activities (see Figure 6.2).

This exercise is actually very important to the estimating function, although again slightly outside the scope of this book. The reader is therefore strongly encouraged to read appropriate texts on project planning and control (see the bibliography at the end of this book).

When a client or their consultants request a programme at tender stage, the contractor will submit a preliminary or outline programme, such as the example given in section 1.8.1. The contractor is often unclear about the role of such a programme in the evaluation of tenders, although they can be extremely useful in assessing whether the tendering contractor has sufficiently engaged with the details of the project. Contractors have also used this opportunity to offer completion sooner than formally required and thereby try to gain an advantage over the competition; however, great care should be taken with this strategy as the client-imposed contract period is often shorter than the optimum duration for normal work productivity, so unilaterally reducing that period for purely marketing purposes may create further difficulties in the construction stage.

6.3.1 Resource histogram and resource levelling

When the estimator has established the labour and plant requirements for the work, and determined how long they will be required for, this information can be added to the tender programme by a resource histogram, as shown in Figure 6.2.

This now gives a graphical illustration of the amount of labour required on site at any one time, which will determine such items as the size of welfare facilities required (canteens, toilets etc.) and also indicate how much project supervision will be required. As it can be expensive for a contractor to move staff around the various sites, they will try to keep the histogram as smooth as possible, while still obviously recognising that more work requires more people. The reader is again referred to textbooks on project planning and control for a discussion of the techniques of resource smoothing and levelling.

6.4 Temporary works

As mentioned previously, temporary works, by definition, do not form part of the finished project and will be dismantled during the course of the construction of the permanent works. A good list of what temporary works would be included in a construction project is shown below, from the UK's Health and Safety Executive website (www.hse.gov.uk).

A. *Simple and/or potentially low risk temporary works*
   - Standard scaffolding
   - Formwork less than 1.2 m high

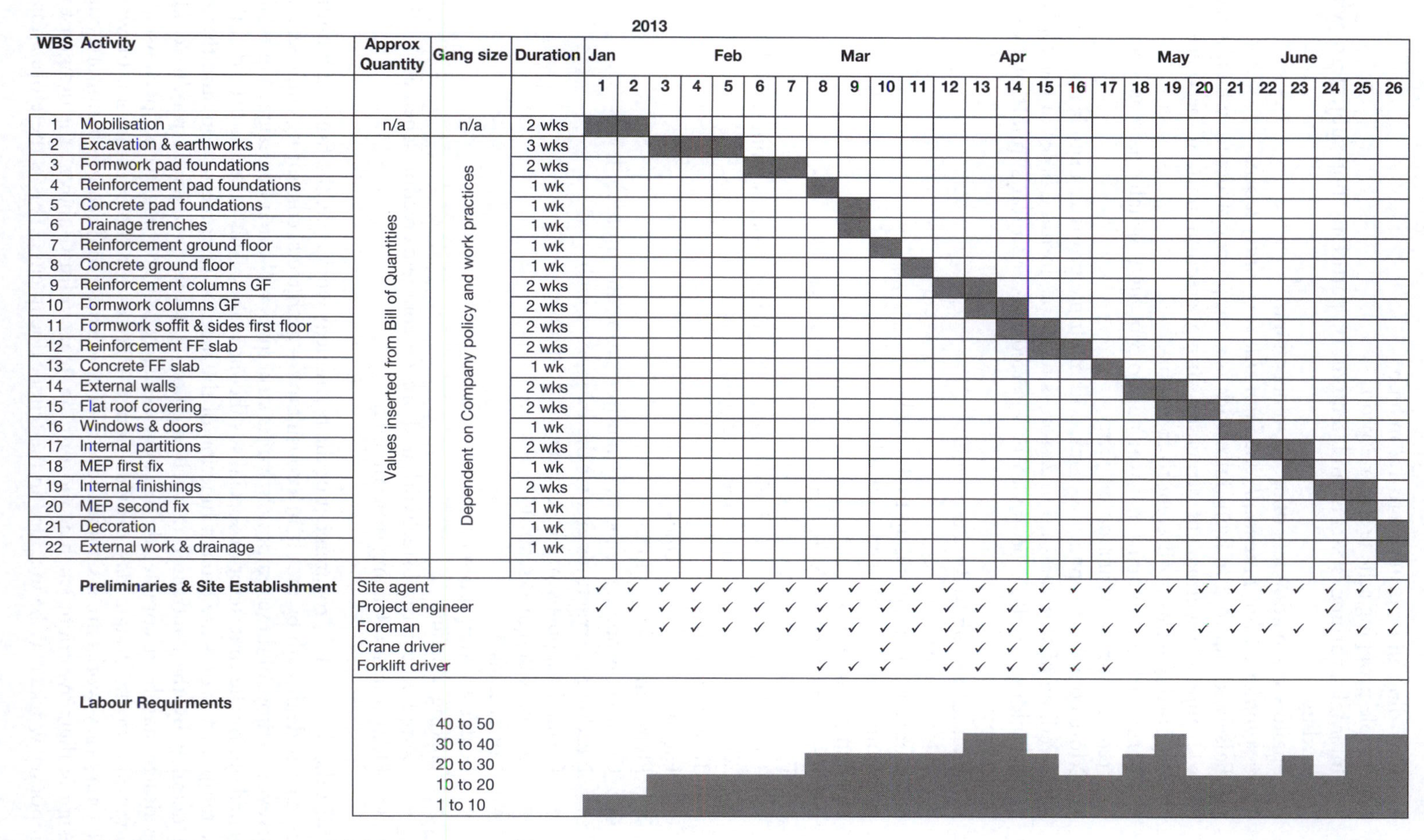

Figure 6.2 Tender programme with labour resource histogram

- Hoarding and fencing up to 1.2 m high
- Simple propping schemes 1 or 2 props
- Internal hoarding systems and temporary partitions not subject to wind loading
- Shallow excavation less than 1.2 m deep/high

B. *More complex and/or potentially medium risk temporary works*
- Falsework up to 3 m high (falsework is a temporary structure which is used to support a permanent structure during erection until it becomes self-supporting – this is different to formwork)
- Formwork for columns and walls up to 3 m high (usually paid as a separate item in most construction projects)
- More complex propping systems, with multiple props at a single level
- Needling of structures up to 2 storeys high (i.e. forming holes in the walls of a building to allow beams to be inserted for structural support of existing floors)
- Excavations up to 3m deep/high
- Safety net system
- Hoarding and fencing up to 3m high
- Simple ad-hoc designed scaffolding
- Temporary roofs

C. *Complex and potentially high risk temporary works*
- Falsework and formwork over 3m high
- Trenchless construction, including headings, thrust bores, mini tunnels
- Working platforms for cranes and piling rigs
- Tower crane bases
- Façade retention schemes
- Flying and raking shores
- Complex propping schemes – multiple props and multiple levels
- Needling of structures greater than 2 storeys high
- Ground support schemes greater than 3m deep
- Complex designed scaffold
- Cofferdams
- Bridge erection schemes
- Jacking schemes
- Complex structural steelwork and precise concrete erection schemes
- Hoarding and fencing over 3m high

Whilst the design and procurement of the permanent works (i.e. the finished building) is normally dealt with by the client's consultants – architects, consulting engineers and surveyors – most of these temporary works structures will often be designed as well as carried out by the contractor or specialist subcontractors. This is because the details of the temporary works are often a function of the method of construction, and this will be detailed by the contractor in their method statement, which is produced after the design has been completed. In some cases, consulting engineers specialising in temporary works design may provide design services for the contractors, or where the temporary works are a significant part of the project and require major structural design in their own right (such as most of the items in list C above), the design of the temporary works may be included in the tender documents. This is therefore a major advantage of engaging the contractor early in the project, so that their skills in method statements and understanding of what temporary works will be required are available whilst the design decisions are being made concerning the permanent works.

Where a contractor or designer is involved in major temporary works, a senior person responsible for establishing and implementing a 'temporary works procedure' should be appointed during the production of the method statement. According to clause 6.3.1 of BS 5975:2008, the items to be addressed are:

- appointment of a temporary works coordinator (if appropriate)
- on larger sites, appointment of a team of temporary works supervisors
- assessing/ensuring competence of design and site-based staff responsible for temporary works
- preparation of adequate design briefs
- the design (including calculations, sketches, drawings, specifications and design risk assessments)
- where appropriate, the designers' method statement
- independent checking of the design
- issue of design check certificates
- procurement of temporary works materials and equipment (in accordance with the designer's specification)
- site control of erection, use, maintenance and dismantling of temporary works
- checking of erected temporary works (and control of use) to ensure compliance with the design
- the issue of the 'permit to load' and 'permit to dismantle' where required.

In terms of the estimating function, all the resource costs associated with these temporary works will need to be established and included in the estimate. Where they are included will depend on the structure of the pricing documents (BOQs, schedule of rates etc.). If there is a substantial preliminaries section in the pricing document, the temporary works should be included in this section as recommended by the RICS NRM. The costs should be included as both lump sum and time-related charges in order to facilitate more accurate payments during the construction process. This is also the position taken by the Civil Engineering Standard Method of Measurement (CESMM) as temporary works in civil engineering projects can represent a significant proportion of the project costs.

For a very interesting example of how to get it completely wrong when trying to reconcile the permanent works with temporary works, the case of the roof design, redesign and re-redesign of the Sydney Opera House has some salutary lessons for the industry.

The Sydney Opera House was constructed in an area that was originally used for wharfing with a tram and storage barn occupying the majority of the site. A design competition was held which required a performing arts complex consisting of two theatres with a large hall for opera, ballet and large-scale symphony concerts, which could accommodate up to 3,500 people. The winner of the design competition was Jørn Utzon, the Danish architect, and he was therefore awarded the contract to design this world famous and iconic monument.

The construction of the Opera House began in early 1959. Even after four years of feasibility and schematic designs, Utzon continued to modify the design and eventually proposed a defined spherical geometry for the roof vaults of the Opera House. The entire construction process had been divided into three main stages – podium or platform, roof structure, and interior design and finishings. The platform and roof were designed by Utzon himself, at an enormous cost in those days. The construction of the first stage (platform) was confronted with several problems such as exceptionally poor weather and technical difficulties which are not uncommon in brown-field contaminated dockland sites. Additionally, because of political pressure from the government departments to complete the project early, construction commenced before the design had been fully approved, causing a considerable amount of rework on site.

The second construction stage consisted of the main part of the structure, i.e. the roof. Even though a contractor had already been appointed and a roof design had been chosen, it was found that this chosen design could not actually be physically built, so several further designs were tested to make a roof that would be both structurally and economically viable. It was finally decided that the shell structures should be created as sections of a sphere, which allowed the contractors to make the arches in a common mould, although of varying lengths, so that they can be placed next to each other to form a spherical shape. Approximately, 2,400 precast ribs and 4,000 roof panels were made in an on-site factory for the construction process.

This, of course, is all history now and the building remains an iconic structure which defines Sydney in the same way that the Eiffel Tower defines Paris (which incidentally was only meant to be a temporary structure for the Great Exhibition of 1889).

6.5 Plant schedules

The contractor's plant requirements will be established from the method statement and construction programme. In order to calculate how long the items of plant will be required on site, the basic performance requirements will be required (in terms of work completed per hour or day) and taken with the total quantity of work, will give the overall duration that the plant is needed. When multiplied by the all-in rate, this will give the contractor the total cost for the item of plant for the project.

Construction plant and equipment are very expensive both to acquire and maintain, as they are subject to considerable wear and tear and are required to be serviced regularly to ensure that they are both efficient and safe. For these reasons, the contractor's plant manager will need to know well in advance when the items of plant are required on the various company sites. Therefore, a plant schedule for each project would be produced, similar to the example in Figure 6.3.

We have already seen in section 2.5 that the company will make a decision of whether to hire or buy the equipment that they require and once this decision is made, the plant manager will be responsible for ensuring that the appropriate items of plant are available to the projects when required.

6.6 Unit rates

There are many techniques used in the construction industry for building up a unit rate to be inserted into a bill of quantities. Let us firstly consider what is meant by a unit rate.

Figure 6.4 shows an example page from a bill of quantities where the quantities of items included in the finished building are set out in accordance with the rules of a

CONTRACTORS PLANT SCHEDULE

Contract____________________ **Contractor**____________________

Person completing the Schedule__________ **Tel./email**____________________

Date____________________

Ref	Work Operation	Plant item	Date requested	Date required on site	Planned duration on site	Release date	Actual Release date	Notes

Figure 6.3 Contractor's plant schedule

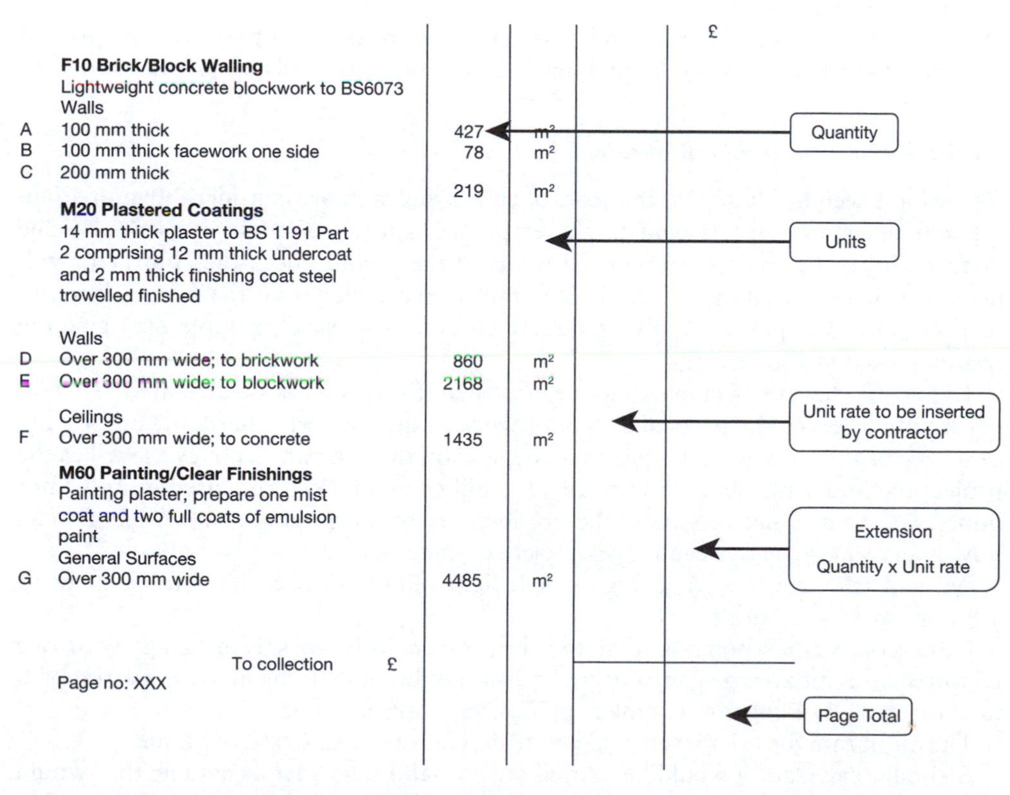

Figure 6.4 Example page from a bill of quantities

standard method of measurement (SMM). As mentioned previously, these quantities are the finished quantities in the project and take no account of wastage of materials, bulking factors etc. These general rules are set out at the beginning of the particular SMM which has been used on the project. As long as the estimator is aware of the rules of measurement, they can make appropriate allowances and be reasonably confident that the estimate has covered all the most important points. Difficulties arise when rough bills of quantities are used with the rules of measurement not being clearly understood by both parties, which is the reason they are called 'standard' methods of measurement.

The contractor's estimator is then expected to calculate a unit rate (i.e. a cost per unit of measurement) for each item in the bill of quantities, which is multiplied by the quantity of the item to arrive at an 'extension cost' for that item. Each page is totalled and carried to a collection/summary, which will be the total estimated cost of the project and this figure is then converted to a tender sum by the addition of head office overheads and profit, and also by being subject to a commercial review by the company directors. See Chapter 7 for a fuller discussion of converting an estimate to a tender.

Most construction projects in the UK, and internationally, use bills of quantities in some form; therefore this is the most common method of tendering construction work. Consequently, a unit rate has to include:

a) the direct labour element for that item of work
b) the cost of the materials needed (with appropriate allowance for shrinkage, bulking and wastage)
c) the cost of any small items of plant which are used solely for that item and are not priced elsewhere, e.g. in the preliminaries section of the bills of quantities.

6.6.1 Labour element of a unit rate

As we have seen in Chapter 2, the costs of employing staff are considerably more than the actual wages that are paid to the employee, and all of these costs need to be factored in to the project estimate. In terms of the labour costs, the contractor will need to calculate a labour 'all-in' hourly rate which includes all the factors listed in section 2.3 in Chapter 2. Applying these factors to a worked example may give the results shown in Figure 6.5.

Therefore, the cost of employing one craftsman for this work is calculated as £9.38 per hour. However, the particular item of work in question may need a team or gang of workers, and the size of this gang will depend on the company's policy as well as the project method statement. Additionally, the bill of quantities is itemised in units, not hours. So, we also need to know the productivity of the person, team or gang – e.g. how many square metres can they complete in one hour?

We will take the item A in Figure 6.4: lightweight concrete blockwork in walls 100mm thick – 427 m^2.

If the contractor's policy is that two bricklayers will work in a 'gang' with one labourer, with an average gang output of 2 m^2 per hour, then the item in question will take the gang 214 hours to complete (427 m^2 at 2 m^2 per hour).

The all-in rate for bricklayers is taken from Figure 6.5 as £9.38 per hour.

A similar calculation would be carried out for labourers – let us assume that would be calculated at £8.50 per hour.

	Item	Cost
(a)	Guaranteed Minimum Wages	11,000.00
(b)	Contractor's Bonus allowance	2,000.00
(c)	Inclement weather allowance	included
(d)	Non-productive overtime costs	350.00
(e)	Sick Pay allowance	100.00
(f)	Trade Supervision	750.00
(g)	Working Rule Agreement allowance	n/a
	SUB-TOTAL (i.e. amounts paid to employee)	**14,200.00**
(h)	CITB training contribution (0.25%)	35.50
(i)	National Insurance contributions	750.00
(j)	Holiday credits	1400.00
(k)	Tool allowance	150.00
	SUB-TOTAL (i.e. Overall costs of employment)	**16,535.50**
(l)	Severance Payment	331.00
(m)	Employer's liability Insurance	333.00
(n)	ANNUAL COST PER OPERATIVE	17,199.50
(o)	NUMBER OF PRODUCTIVE HOURS WORKED	1834
	ALL-IN HOURLY RATE	**9.38**

Figure 6.5 Worked example of calculation of all-in rate for craftsman

The total hourly rate for the 2:1 gang would be £9.38 + £9.38 + £8.50 = £27.26

If it would take 214 gang hours to complete the work, therefore the total labour cost would be: 214 × £27.26 = £5,833.64

Clearly, each company's costs would be different, reflecting the rates they pay to their workforce, the additional on-costs included in the all-in rate, but most crucially, the performance and productivity factors used to calculate the time required for the operation. The more efficiently that work is carried out in an organisation, the lower will be the average costs. Inefficiency itself is a form of wastage and the effect of this wastage is to make the contractor much less competitive and therefore less likely to be successful in tendering.

6.6.2 Material element of a unit rate

As mentioned in section 2.4 of Chapter 2, the cost of materials, which should be relatively similar for all contractors, can vary considerably depending on the design of the works and the efficiency of the site management. For example, the cost of materials can be affected by:

a) the actual quantity required – a larger project will have lower average costs of materials than a smaller project
b) delivery to site in small or part loads, which will incur extra transport costs
c) storage and protection requirements for fragile materials and components, or where security from theft is a concern
d) size and weight of materials and components requiring special site transport and fixing methods
e) subsidiary fixing materials or falsework required during erection
f) wastage factors, which are generally much lower for pre-manufactured and 'dry' components and higher for site-based construction and 'wet' trades
g) costs of complying with environmental legislation, such as COSHH and Site Waste Management Regulations.

Contractors with a structured supply chain management policy with their preferred suppliers will be able to calculate the materials element of the rates relatively easily and quickly. The larger firms will be able to negotiate greater discounts with the suppliers, but will also have greater overheads because of their size. Also, larger firms may engage in 'vertical integration' where they have their own concrete production facility, or in-house MEP abilities etc. All of these issues can affect the cost of materials which would be included in an estimate or tender.

6.6.3 Plant element of a unit rate

As mentioned in section 2.5 of Chapter 2, the costs of plant and equipment can vary greatly depending on the item of plant chosen, its efficiency and productivity together with whether the item is owned by the contractor or hired from a specialist plant hire company. Costs of small items of equipment are either included with the labour element (if the tradesmen owns their own tools, as many of them prefer to do) or they would be included in either the preliminaries or general overheads. Items of plant to be included in the unit rates should be restricted to those which are only to be used for the item in question; otherwise the contractor is in the complex and ridiculous situation of 'sharing' items of plant and equipment between items in the bill of quantities.

Additionally, it has also been noted in Chapter 2 that plant and equipment costs can either be fixed costs (mobilising to site, erecting etc.) or time-related costs (hire charges, operator's wages etc.). Only a relatively small percentage of the plant costs can be considered to be quantity related, i.e. costs which rise depending on the amount of work done by the equipment – an example would be fuel costs. However, the items in the bills of quantities are, by definition, measured in terms of the quantity of work done, so any plant costs to be allocated to the unit rates must be converted to this format.

Again, we will take the example of item A in Figure 6.4: lightweight concrete blockwork in walls 100mm thick – 427 m^2.

The contractor has decided to mix the mortar for this blockwork by using a small electric-powered cement mixer with a capacity of 0.3 m^3.

We have seen in section 6.6.1 that it would take 214 hours to complete this item of work.

If the all-in rate for this item of plant is £4.00 per hour, the total plant element for this item is £856.00, which is a very high figure given that we could buy one of these machines for around £3,000 with a working life of several years. Clearly, there is little need for the item of plant to be allocated totally to this rate and should also be used by other bill of quantities items requiring mortar, such as items B and C.

6.6.4 Contractor's extended bills of quantities

When all the build-ups to the rates have been calculated, the tendering contractor will be in a position to establish labour, materials and plant elements for each item in the bills of quantities supplied with the tender documents. As most bills of quantities are now supplied in electronic format, using standard spreadsheet software such as Microsoft Excel, it is relatively easy for the estimator to add extra columns to the spreadsheet to insert the labour, materials and plant costs of each item (see Figure 6.6 for an example). If the item is predominantly covered by a subcontractor, this can be added in an additional column. All of these elements will add up to the item rate, which will be inserted in the bill of quantities to be submitted with the tender, if required.

Each column, for labour, materials and plant will produce a total for that element of the rate, which will be carried forward to the appropriate summary sheets, as described in section 7.1 and Tables 7.1, 7.2 and 7.3.

Example of a page from a Bill of Quantities			Labour	Materials	Plant	S/C	Rate	£
F10 Brick/Block Walling								
Lightweight concrete blockwork to BS6073								
Walls								
A 100 mm thick	427	m^2						
B 100 mm thick facework one side	78	m^2						
C 200 mm thick	219	m^2						
M20 Plastered Coatings								
14 mm thick plaster to BS 1191 Part 2 comprising 12 mm thick undercoat and 2 mm thick finishing coat steel trowelled finished								
Walls								
D Over 300 mm wide; to brickwork	860	m^2						
E Over 300 mm wide; to blockwork	2168	m^2						
Ceilings								
F Over 300 mm wide; to concrete	1435	m^2						
M60 Painting/Clear Finishings								
Painting plaster; prepare one mist coat and two full coats of emulsion paint								
General Surfaces								
G Over 300 mm wide	4485	m^2						
To collection £								
Page no: XXX								

Figure 6.6 Contractor's extended bill of quantities

6.7 Subcontractors

Most main contractors use other firms to carry out some of the works on a project. Again, as we saw in section 2.6 in chapter 2, if the main contractor chooses to subcontract the works themselves (often with required approval from the client), these are known as 'domestic' subcontractors and there is no differentiation in the bills of quantities – the rates are priced as though the main contractor carried out the work, and indeed they still retain the responsibility for this element of the works. If the works are carried out by a named or nominated subcontractor, the relevant work would be covered by a prime cost (PC) or provisional sum – see section 6.8.

Domestic subcontractors would therefore provide a quotation to the main contractor, which is confidential between the two companies. This quotation would act as the 'prime costs' for the main contractor for these items – i.e. the allowance for labour, plant and materials in the unit rate to be inserted in the bill of quantities.

In addition to these prime costs, the main contractor would need to consider the following possible extra costs for domestic subcontractors:

1. Has the subcontractor quoted for everything that was in the main contractor's enquiry and also in the latest scope of works from the client or consultants?
2. Are there any special requirements for unloading, storage and protection of domestic subcontractor's materials and components? Domestic subcontractors are chosen because of the specialist nature of their expertise. Therefore, the materials they are required to work with may be delicate, fragile or expensive.
3. What general attendance requirements does the domestic subcontractor require? Their work may be very intensive over a short period of time; therefore the labour requirements on site may be high for this short time period. Can the main contractor's existing welfare facilities cope with this extra burden?
4. Are there any other special attendances required, such as three phase electrical supply?

The Code of Estimating Practice gives three options for the main contractor to make these appropriate allowances:

- The relevant unit rates for the subcontracted work may be increased to cover for 'main contractor's overheads'.
- A fixed percentage can be added to the subcontractor's quotation and either distributed across the relevant rates or added to appropriate items in the preliminaries section.
- The general project overheads can be increased.

Many domestic subcontractors offer discounts to the main contractor (hopefully in anticipation of being paid on time) and unless the main contractor is particularly keen on submitting an ultra competitive bid, these discounts are unlikely to be passed on to the client and should be accounted for in the main contractor's profit element.

Finally, the definition of a subcontractor is one who provides a service and this normally consists of provision of labour and materials (and maybe equipment). However, it is possible to provide a 'labour-only' subcontract, which effectively means that the subcontractor is a self-employed tradesman who does not provide any of the materials or equipment needed for the project. There are very strict government regulations in the UK concerning labour-only subcontracting and all such subcontractors must be registered with the tax authorities to ensure that the proper amount of income tax is paid. In this case, the main contractor needs to ensure that all the appropriate costs are covered in the project estimate – see also section 9.3.2 in Chapter 9.

6.8 Provisional sums and dayworks

The Code of Estimating Practice includes a flowchart to illustrate the structure and purpose of PC and Provisional Sums – see Figure 6.7.

This section relates to only provisional sums and dayworks (i.e. the right-hand side of the chart in Figure 6.7) as the main standard forms of contract in the UK have deleted the entire nominations system due to legal difficulties encountered in the early 2000s with the process of nomination. However, many other countries across the world still operate a process of client nomination of subcontractors.

The quotation from the accepted nominated supplier or nominated subcontractor would not be available to the main contractor until after the main contract has been awarded; therefore, they are required to put an allowance in their tender for this quotation. This allowance is covered by the PC sum and the main contractor is allowed to add their required profit margin and any costs for general and special attendances. The definition of these attendances is given in section 2.6.2. In accordance with SMM7 rules of measurement, the main contractor's profit should be given as a percentage but the general and special attendances should be priced in detail as an item. The NRM is silent on this issue as all these costs would be covered by the provisional sum.

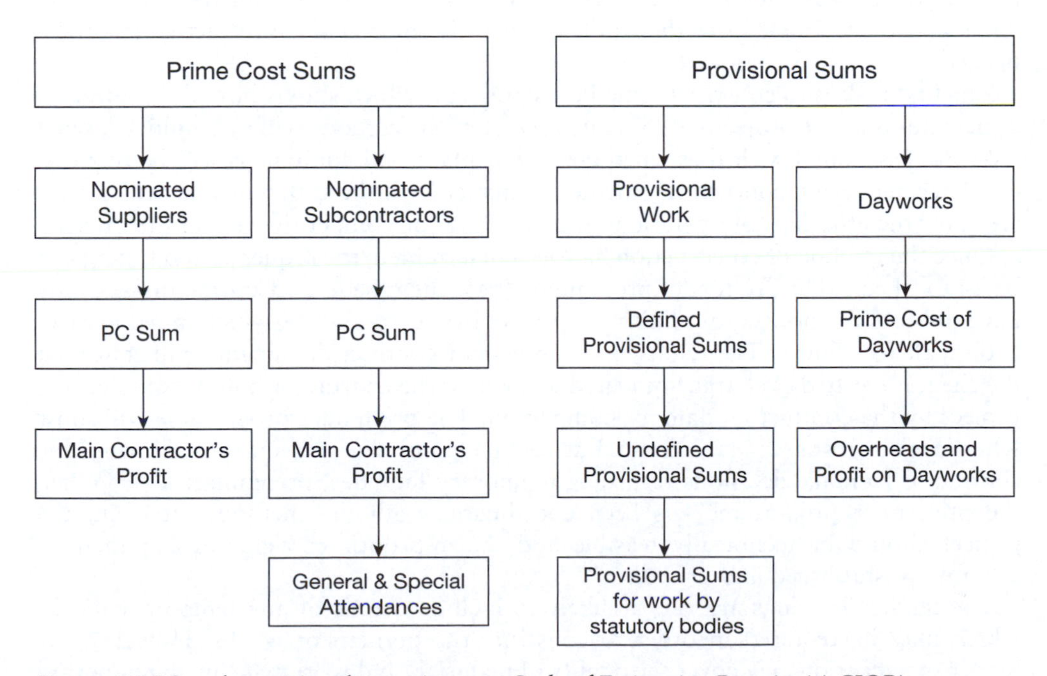

Figure 6.7 PC and provisional sums (source: *Code of Estimating Practice* (c) CIOB)

The right-hand side of the flowchart separates provisional sums into sums covering provisional work and sums covering dayworks. Provisional work is work that is known to be required but cannot yet be defined in detail – see section 2.6.3 and note particularly the difference between defined and undefined provisional sums as this could add significant extra costs to the client. Work required to be carried out by statutory authorities, such as electrical, gas or sewerage connections, are also covered by provisional sums. Section 2.6.3 also gives a definition of dayworks and a provisional sum will be added in this section for both the actual cost of labour, materials and equipment used (covered by approved timesheets and invoices) together with a separate allowance for main contractor's overheads and profit on dayworks – which could be as high as 200 per cent due to the disruption caused to the contractor's day-to-day management.

6.9 Summary and tutorial questions

6.9.1 Summary

When the design is complete, the project scope is known. This of course is obvious and self-evident, but it also means that the work can be estimated and priced with much greater certainty and the resources that are required for the project can be established along with both the method and scheduling of construction operations.

If the contractor (builder) has not been involved in the design stage, they will normally receive a set of drawings representing the proposed finished works together with the specification of materials and workmanship required on the project. They will normally be expected to submit a lump sum price to complete this work in the time period stated in the tender documents. They should now have all the information necessary to accurately price the works using anticipated costs of the actual resources needed.

The first task is to construct a method statement, which shows how the contractor anticipates that the project will be constructed, what methods will be employed, what risks are associated with these methods, what plant and equipment will be required etc. Each tendering contractor's method statement will be different and will set the scene for the efficiency of operations and therefore the competitiveness of their tender.

Once the method of construction has been determined, the sequence and timing can be established in the pre-tender programme (baseline schedule). Contractors not only have to build the project, but finish the project by a certain date, which in many cases is difficult to achieve. The tender and subsequent contract documents will state that the client is entitled to charge liquidated damages if the contractor fails to complete the project by the contractual date for completion. The pre-tender programme will show what level of resources are required at each stage of the project, thus allowing the contractor to make decisions regarding temporary facilities, preliminaries etc. When the pre-tender programme has been coordinated with the method statement, the project should be technically feasible and the appropriate resources can then be accurately established and priced.

The tender drawings are very unlikely to include details of any temporary works which may be required in order to construct the permanent works. However, the tender specification may cover the quality of materials and workmanship in temporary works, to ensure that any temporary works are robust enough for the project requirements. Temporary works are generally at the risk of the contractor and there is usually no facility to separately price them in a bill of quantities (unless it is a civil engineering project and the bill has been measured in accordance with CESMM4, where temporary works can be measured and itemised separately). Therefore the cost of temporary works must be spread across the rates for the permanent works, or included in the project preliminaries.

A bill of quantities is normally structured in trades or work sections in accordance with either of the lists given in Table 5.5 in Chapter 5. The quantities relate to the finished quantities in the completed building and the unit rates which the tendering contractor is to insert are the rates for one unit of the item in question. As the contractor has built up their costs from first principles (i.e. the cost of the actual resources – labour, materials and plant), they must convert these costs to unit rates for inclusion in the bill of quantities. Therefore each unit rate will include an element for labour, an element for materials and an element for any plant and equipment which is not included elsewhere. How these are calculated and incorporated into the unit rate is explained in section 6.6.

The remaining costs represent the work carried out by subcontractors or those project costs which cannot be established at the tender stage. Domestic subcontractor's costs will form part of the normal unit rates as the main contractor still has the responsibility for this work. Any nominated or named subcontractors will normally be covered by a PC sum or provisional sum and the costs allocated to these items will be adjusted and confirmed during the construction stage when the appropriate subcontract scope of works is confirmed.

Completing the final costs of a construction project will be the preliminaries. Section 2.7 in Chapter 2 discusses the costs associated with preliminaries and general site overheads, all of which will need to be firmly established once the method statement and tender programme have been agreed.

6.9.2 Tutorial questions

1 What is the purpose of a contractor's resource histogram and how will it impact on the pricing of preliminaries?
2 Discuss the various alternatives available to a tendering contractor to include the costs of temporary works in a tender.
3 What are the advantages and disadvantages to a tendering contractor of structuring a bill of quantities in accordance with:
   a) building elements
   b) work sections?
   Refer to Table 5.5 in Chapter 5.
4 Comment on the cost significance of the difference between a defined and undefined provisional sum.
5 What are the other uses of a unit rate apart from helping to build up to a lump sum tender?

7 Conversion of a contractor's estimate

When the project estimate has been completed, or probably more realistically, when the tender submission date is getting closer and no further extensions of time can be squeezed out of the client, the raw estimate produced by the estimating department must be converted to an actual tender price which will be submitted to the client in competition with all other tenderers on the project.

For the projects where there is full design information available in the tender documents, the tenderers will have established the following:

a) the method that will be adopted for constructing the building (method statement)
b) the programme that will be adopted showing when each operation in the method statement will be carried out on site, with appropriate linkages to associated operations (pre-tender programme)
c) the anticipated actual cost of labour required for this project, or more specifically, the actual cost of labour to the company during the current year, if the all-in rates are calculated on an annual basis
d) the anticipated actual cost of the plant and equipment required for this project, although small items of plant and equipment will be included within the contractor's general overheads (as they are not regarded as items of capital expenditure)
e) the anticipated actual cost of materials required for the project, taking into account any discounts, bulk purchases, allowance for wastage and transport costs to the site.

The costs in (c) to (e) above are part of the direct costs of the project (see Chapter 2) and will be the costs which are actually paid out by the contractor during the construction operations. It is useful to note here when these costs will be paid out – the labour costs will predominantly be paid out on a weekly basis to the site operatives, although some supervision and office staff will be paid on a monthly basis. The plant and equipment costs will be paid depending on who owns the equipment; if it is hired from a specialist company, then the costs will probably be paid monthly or in lump sums; if it is directly owned by the contractor, the costs will represent repair and maintenance etc. which is an ongoing cost, as well as depreciation which does not actually involve cash payments anyway. Material costs will be paid on account to suppliers, which are often paid on terms ranging from 30 days to 90 days if they are lucky.

The point that we are making here is a reiteration of the point made in Chapter 6, that costs go out on a weekly, monthly and quarterly basis, but the project payments come in up to two months after they are incurred, which consequently creates a constant negative cash flow for the contractor, especially if retention is applied to their payments from the client. The senior managers of the tendering contractors will need to take account of this when converting their raw estimate to a competitive tender.

7.1 Management summary and estimator's report

The direct project costs mentioned above will be structured in terms of the unit rates for all the items in the bill of quantities. However, at the tender submission stage, the contractor's senior management will require to be provided with a summary of all the labour costs, plant costs and materials costs in the project. These summaries will be produced by the estimating department in the formats shown in Tables 7.1 to 7.3.

In Table 7.1, the total amount of labour costs which have been included in the BOQ is shown at (A) and split between the company's own employees (B) and any labour-only subcontractors (C). In this way the company management can readily see how much of their company's employed labour costs will be covered by this contract (if they win it, of course). However, there may be some adjustments to be made based on, say, further discounts from the labour-only subcontractors or late quotations from other firms which will have the effect of reducing the overall costs. In this case, the labour costs to the main contractor would be lower overall, but it may be too late to change the rates in the bill of quantities, so either the contractor's profit margins would rise, or an adjustment could be made to the spread percentage if it is thought necessary to make the project tender price as competitive as possible.

Table 7.2 shows the summary of plant and equipment costs for the project. All the various items of major plant are described, including the productivity factors used in the calculations, which, when multiplied by the total quantity will give the required duration of the equipment on site. This duration multiplied by the rate per hour will then give the total cost as used in the raw estimate. The project planner will have created a plant and equipment programme showing all the items of plant on site at any one time, which is clearly necessary for the site logistics of ensuring the items of plant do not interfere with each other and also allows decisions to be made regarding where to store the items of equipment overnight. The total cost of the plant and equipment is given in column (E). Again, if late information is received which increases or reduces the overall cost of the plant, this will be shown in the last three columns and totalled in column (F).

Table 7.3 shows a summary of material costs with the total BOQ amount given at column (G) and adjusted total given at column (H).

7.1.1 Allowance for inflation and increased costs

The vast majority of modern construction projects are let on a 'fixed price' basis, therefore any increase in prices or inflation of the costs of materials or labour costs will be the responsibility of the contractor. This is not usually considered a major problem when inflation is low but would be a significant risk during periods of high inflation, as was experienced in the 1970s and 1980s when inflation peaked at almost 18 per cent per annum. Most standard forms of construction contract include an optional set of clauses which allow the contractor to reclaim inflation costs (called 'fluctuations') and in order to prevent the mind-numbing task of dealing with literally thousands of

Table 7.1 Summary of project labour costs

	XYZ Construction Ltd			**LABOUR SUMMARY**			Project:	OMR Residential Housing	
							Tender No.	RT4326	
							Date of Tender:	September 2012	
Ref	Description	Trade	Total hours	All-in rate	Total £ in BOQ	Company employees	Labour only S/C	Adjustments	Adjusted total
	TOTALS				**A**	**B**	**C**		**D**

Table 7.2 Summary of project plant and equipment costs

	XYZ Construction Ltd			PLANT SUMMARY					Project:	OMR Residential Housing	
									Tender No.	RT4326	
									Date of Tender:	September 2012	
Ref	Description of item of plant	Quantity	Unit	Output used in estimate	Duration	Rate/hr	Total cost	Sub-contractor	Quote	Adjustments	Adjusted total
	TOTALS						E				F

Table 7.3 Summary of project materials costs

<table>
<tr><td></td><td colspan="2">XYZ Construction Ltd</td><td></td><td colspan="3">MATERIALS SUMMARY</td><td></td><td>Project:</td><td colspan="2">OMR Residential Housing</td></tr>
<tr><td></td><td></td><td></td><td></td><td></td><td></td><td></td><td></td><td>Tender No.</td><td>RT4326</td><td></td></tr>
<tr><td></td><td></td><td></td><td></td><td></td><td></td><td></td><td></td><td>Date of Tender:</td><td colspan="2">September 2012</td></tr>
<tr><td></td><td></td><td colspan="3"></td><td></td><td></td><td></td><td></td><td></td><td></td></tr>
<tr><td>Ref</td><td>Material description</td><td>Supplier</td><td>Quantity from BOQ</td><td>Unit</td><td>Wastage factor</td><td>Total quantity</td><td>Net rate</td><td>Total in rates</td><td>Adjustment</td><td>Adjusted total</td></tr>
<tr><td></td><td></td><td></td><td></td><td></td><td></td><td></td><td></td><td></td><td></td><td></td></tr>
<tr><td></td><td></td><td></td><td></td><td></td><td></td><td></td><td></td><td></td><td></td><td></td></tr>
<tr><td></td><td></td><td></td><td></td><td></td><td></td><td></td><td></td><td></td><td></td><td></td></tr>
<tr><td></td><td></td><td></td><td></td><td></td><td></td><td></td><td></td><td></td><td></td><td></td></tr>
<tr><td></td><td></td><td></td><td></td><td></td><td></td><td></td><td></td><td></td><td></td><td></td></tr>
<tr><td></td><td>TOTALS</td><td></td><td></td><td></td><td></td><td></td><td></td><td>G</td><td></td><td>H</td></tr>
</table>

invoices from suppliers, the contractor would usually be paid by means of a formula method based on the index of construction costs published by reputable organisations such as BCIS or the relevant government department of national statistics. The base index would usually be set one month before construction starts and the contractor would be paid for fluctuations based on the rise in this index number throughout the construction period on a reducing balance basis.

7.1.2 Risk analysis

As mentioned in a previous chapter, all proposed activity in the future carries some degree of uncertainty. Risk is another word for uncertainty. When we buy a lottery ticket, we are uncertain about whether it will win (in fact the probability of winning the jackpot in the UK national lottery is over 14 million to 1, so we can say that there is a high probability that we are not going to win) therefore, we are risking the cost of the ticket against the probability of winning.

The CIOB Code of Estimating Practice classifies construction risks into two groups:

a) *'The Known-Unknowns'*, these are uncertainties that the contractor can make allowance for since they will affect the cost of the construction but not in a major significant way. Examples are health and safety, environmental issues, design completion, construction difficulties, duration changes due to client-imposed variations, existing ground conditions, previous experience of working with the client and consultants, terms and conditions of contract.
b) *'The Unknown-Unknowns'*, these would be issues related for force majeure or 'Acts of God', which if they occurred would have a catastrophic impact on the project. Examples would be major floods, riots, civil commotion etc. One could say that in the twenty-first century then major floods and natural disasters occur all too frequently and the civil unrest which characterised the year 2011 in many parts of the world means that this risk needs to be taken very seriously and factored in to the project costs.

The contractor should therefore develop a 'risk register' where all the anticipated risks are itemised if it is felt that they could have a financial effect on the project and thereby reduce the contractor's profit margin.

These risks will be assessed by the contractor for the severity of impact if they were to occur, together with the probability of their occurrence. Table 7.4 illustrates what should be done with this assessment.

Issues in the boxes entitled low or moderate risk can normally be safely managed by the company through its own day-to-day site or office procedures or if there is a financial loss, it will be relatively small and is unlikely to do significant harm to the profitability of the project. Issues in the boxes entitled high risk will need to have a strategy put in place to deal with the risk that has been highlighted, as any occurrence will have a significant physical and/or financial effect on the project. Examples of strategies that could be adopted include:

a) *Risk avoidance* techniques such as changes to design or method which will avoid the hazard.

Table 7.4 Risk matrix

		CONSEQUENCES				
		Insignificant	**Minor**	**Moderate**	**Major**	**Catastrophic**
LIKELIHOOD		Minor problems easily handled by normal day-to-day processes	Some disruption possible and financial cost	Significant time and resources required to deal with issue	Operations severely affected with major cost impact	Business survival is at risk
	Almost certain to occur	**HIGH**	**HIGH**	**EXTREME**	**EXTREME**	**EXTREME**
	Very likely to occur	**MODERATE**	**HIGH**	**HIGH**	**EXTREME**	**EXTREME**
	Moderate chance of occurring	**LOW**	**MODERATE**	**HIGH**	**EXTREME**	**EXTREME**
	Unlikely to occur during project	**LOW**	**LOW**	**MODERATE**	**HIGH**	**EXTREME**
	Very rare	**LOW**	**LOW**	**MODERATE**	**HIGH**	**HIGH**

b) *Risk mitigation* techniques where measures are taken to reduce the impact on the project, for example increasing the security on a project in areas of high crime, or a risk of civil commotion. For this case, it may be prudent for the contractor to allow cost and time contingencies in their estimate and programme.
c) *Risk transfer* techniques where a risk is transferred to another party who may be better able to carry the risk due to their technical expertise and specialisation.
d) *Insurance* against the risk issue if this is possible. This is a form of risk transfer although additional costs will be incurred.

Strategies (a) and (b) are internal to the project as the contractor seeks to manage the risk themselves, whereas (c) and (d) are external as the risks are transferred to other parties or companies. Whatever strategy is adopted, it should be well considered and thought through at tender stage, for all the hazards and risks that are foreseeable at the time. One of the basic characteristics of modern construction procurement techniques is that more of the risks in projects are transferred from the client to the contractor. Traditionally, the contractor would construct the building based on a full design issued by the client's consultants, the contractor therefore having no design risk and the terms and conditions of the traditional forms of contract allocated the remaining risks reasonably fairly between the parties. The modern trend towards package deals, turnkey, design-build, EPC and PFI contracts significantly changes that landscape, with the contractor being required to take on much more of the technical and in some cases the economic and financial risk of the project.

Fortunately, construction estimators are well trained and experienced in foretelling the future and therefore make good risk managers. This is due to their ability to break down the project into its constituent parts during the estimating process, and they also possess well-developed analytical and quantification skills to be able to develop a comprehensive risk register at tender stage.

7.1.3 Estimator's report

When the estimate has been completed and is ready for the senior management review prior to being submitted to the client as a formal tender, the estimator will submit a report to the senior management highlighting exactly what has been allowed for in the estimate. The Code of Estimating Practice recommends that this report should include the following sections:

a) *The project and its characteristics.* The report should start with a general introduction highlighting the size, nature and overall construction of the building including any special technical or commercial features which should to be brought to the senior manager's attention at the tender stage.
b) *Procurement details and contract conditions.* As there is such a wide variety of procurement routes available to clients in the modern construction industry, the directors of the contracting firm will require to know exactly how this project has been structured in order to ensure that there are sufficient risk allowances in the tender. Additionally, they will also wish to know if a standard form of contract has been used including any amendments to the standard terms and conditions. If a standard form has not been used and the conditions of contract are on an ad-hoc basis, the company will require a full and proper legal assessment of the

conditions in order to establish the degree of risk to which the contractor will be exposed.

c) *Method statement.* The full method statement should be included in the report ensuring that it fully describes the sequence of work, the interface dates between phases and subcontractor involvement together with a summary of resources (more detail given in the summary tables) and construction logic, which has formed the basis of the pre-tender programme.
d) *Pre-tender programme.* The full programme schedule developed from the method statement by the project planners should be included, setting out all the activities, durations and construction logic. The critical path should be clearly marked. The programme should have been prepared from the latest version of an industry standard programming software package, such as Primavera.
e) *Special site factors.* In addition to the general characteristics mentioned in (a) above, any special site factors which would have an effect on logistics or incur special costs should be mentioned separately. Examples would include any restrictions on access and egress from the site (narrow roads, no deliveries during 'rush hour' etc.) and any constraints on normal operating procedures (overhead lines restricting tower cranes etc.)
f) *Subcontractors and suppliers.* Details should be given in this section regarding which companies are intended to be used in the supply chain for this project. Many main contractors have long-term agreements with their supply chain partners and it is therefore expected that these firms will be appointed. Clients may also insist on providing materials free of charge from either their own facilities or one of their subsidiary companies, which should also be noted in this section.
g) *Preliminary costs.* As stated previously, preliminary costs relate to the contractor's overheads which occur on site and include all site establishment costs, supervision and general items of plant and equipment which cannot be allocated to specific items in the permanent work. These costs therefore reflect the efficiency with which the project is managed on site and the directors will have a good idea of the normal percentage for preliminaries for a project of this nature from the company's past experience.
h) *Health, safety and environment (HSE) issues.* Given the important statutory obligations in this area together with the negative implications of poor HSE performance, the critical issues relevant to this particular site will need to be identified so that they can be managed as effectively as possible. Of particular concern will be working at height, heavy loads and proximity of adjoining properties. Adjustments to the method statement can reduce the impact of all these potential hazards.
i) *Cash flow.* Based on the pre-tender programme and the estimated costs of the works together with the payment terms in the conditions of contract, an expected cash flow should be produced to allow the directors to assess whether or not business financing will be required for the project, or whether the project will contribute to the business cash flow throughout the construction stage. All businesses run on cash flow and the company will have many other projects which are all at different stages of their life cycle. Therefore, an assessment of how this project will contribute to the overall business cash flow is vital for good corporate financial management.

j) *Bonds, warranties and guarantees.* Purchasing bonds and warranties from a bank or financial institution can be an expensive and time-consuming business, therefore the contractual requirement for such bonds should be highlighted to the directors as soon as possible. These costs are normally covered in the project preliminaries, as project-related indirect costs
k) *Insurances.* Most contractors carry their own Contractor's All Risks (CAR) Insurance covering the main standard items which they are required to insure against. If the particular project requires further or specialist insurance, this should be highlighted. Again, these costs are normally covered in the project preliminaries.
l) *Client risk analysis.* As the contractor will be taking a significant financial risk on any construction project, it is advisable for them to assess the client organisation in terms of their position in the marketplace and historical performance with all their previous construction projects, especially regarding prompt payments and any propensity to interfere in the projects on site. This analysis is critical in projects such as PFI where the 'client' may only be a shell company set up for a particular development and may have no assets or background of their own. In these circumstances, the contractor may be well advised to seek a payment guarantee bond from the client's bankers or a parent company guarantee from the more substantive companies who are shareholders to the shell company. This is a very tricky area in the modern construction industry.
m) *Consultant risk analysis.* The consultants forming the design and supervision teams should also be assessed in the same way as (l) above, especially regarding their size, capability and experience for the project in question. Working with inexperienced or incompetent/under-resourced consultants and designers can significantly increase the risks and costs to a contractor.
n) *Design and quality expectation.* This also relates to the design consultant analysis in (m) above in terms of what is expected in terms of the quality of construction and finishings. Clearly, a five star hotel will require high-quality finishings, but are the same standards of quality expected for a two star hotel? If so, this may say something of the potential relationships during the construction stage.

The company management would now have a clear picture of the total estimated cost of the project, split down into the constituent elements of labour, plant and materials. As stated in Chapter 2, the raw estimates, or base costs, are unlikely to be significantly different from other tenderers (except for methods of working and any errors or omissions), so the competitive element of head office overheads and profit need to be added to the estimate in order to convert it to a tender figure, which will hopefully be competitive enough to win the project, or at least be invited for further negotiations.

7.2 Converting an estimate to a tender

All the total project costs of labour, plant, materials, subcontractors, site-based overheads (preliminaries) and provisional sums can now be shown on one spreadsheet, an example of which is shown in Table 7.5.

Even though the costs have undoubtedly been calculated as accurately as possible by the estimating department, the directors and senior managers will take a global view of the project including a comparison of any benchmark figures with other similar projects in order to try to reduce the costs in each element in an attempt to make the

Table 7.5 Tender summary of project

	XYZ CONSTRUCTION LTD			TENDER SUMMARY			Project:	OMR Residential Housing	
							Tender No.	RT4326	
							Date of Tender:	September 2012	
		From Estimate							
		Labour	Plant	Materials	Sub-contractors	Prelims/site overheads	Provisional sums	Sub-total	TOTAL
1	Own measured work								
	Direct work	40,000	87,000	720,000				847,000	
	Domestic SC & attendances	3,000	2,000	2,400	2,520,000			2,527,400	3,374,400
2	Preliminaries/site overheads							–	
	Fixed costs	24,000	16,300	5,000	400	5,600		51,300	
	Time-related costs	40,000	102,000	6,300	200	132,000		280,500	
	Insurances					10,000		10,000	
	Bonds & guarantees					15,000		15,000	356,800
3	Provisional sums – defined						250,000	250,000	250,000
4	Adjustments							–	
	Firm price			25,000	36,000	4,000		65,000	65,000
	Final review	–4,600	–6,500	–56,000	–55,000	–23,000		–145,100	–145,100

	Total Net Costs	102,400	200,800	702,700	2,501,600	143,600	250,000	3,901,100	3,901,100
				Mark-up required for:					
				Scope definition		nil	%		
				Risk factors		3	%	117,033	
				Head Office overheads		5.50	%	214,561	
				Profit		3.50	%	136,539	
				Sub-total					468,132
				Dayworks - Undefined Provisional Sum					150,000
				Other Undefined Provisional Sums					nil
				TENDER TOTAL EXCLUDING VAT					£ 4,519,232

project price more competitive. In this case, should the project be successfully awarded, the construction team will be required to make the appropriate savings on site, in order to ensure that budgets are maintained.

It should also be noted that, on occasions, tendering contractors may decide that they do not want to be awarded a particular project but also do not wish to forward a letter of regret for fear of being excluded from the tender list of future projects. They will then be tempted to submit a tender price which is very high in the expectation that they will be unsuccessful. This high tender price is called a 'cover price' and this practice is now illegal in the UK and many other jurisdictions under competition law and anti-trust legislation. The Office of Fair Trading (OFT), part of the UK government, imposed some considerable fines on various firms for this practice in 2009. The OFT argued that for a contractor to know that a tender price is high enough to guarantee that the tender will be unsuccessful means that they must know what the other tenders are going to be, which is therefore collusion and consequently anti-competitive and illegal. Of course all collusion is unacceptable in a competitive environment, but the argument also fails to recognise that the contractors will know what the real 'base costs' of the project are likely to be and will also know what margins are likely to be successful in other contemporary projects, as part of their market analysis. Therefore, submitting a 'cover price' may not involve collusion at all, it can be a product of good market intelligence and market analysis – both of which should be encouraged as good business practice.

7.2.1 The settlement meeting

The settlement meeting will be the point at which the estimate is officially converted to a tender and approved by the senior management for submission to the client on the required date and time as stated in the tender instructions. The technical and commercial matters highlighted in the estimator's report will be discussed and checked to ensure that they are valid and sufficient and any assumptions are acceptable. Naturally, all procedures should have followed standard company practices, which will have been developed for the purpose of minimising any risks of errors and omissions.

The mark-up shown on the final summary form (see Table 7.5) will include the allowance that the company has made for all risks, head office overheads and required profit. Depending on the contract conditions, undefined provisional sums will not have had this mark-up allocated, as they will be allowed to include a mark-up percentage by virtue of being undefined at tender stage. The mark-up percentage can also be known as the 'spread' percentage.

Other decisions which need to be made at the settlement meeting include:

a) What documents are to be submitted with the tender? These should be specifically established in the tender documents from the client.
b) Has an alternative tender been requested? i.e. with contractor alternatives offered which may benefit the client both technically and commercially.
c) Has the tender been qualified in any way which may result in it being rejected?
d) Do the tender documents require a fully priced bill of quantities, or is this only required if the tender is being seriously considered for acceptance?

The settlement meeting will review all aspects of the bid and should have formal minutes of the meeting, setting down all decisions reached. The submission of the tender is the responsibility of the senior management of the company; the estimator's responsibility is to inform them of the likely costs of the project and this will be handed over to the bid manager/director to make the final commercial decisions and hopefully secure the work for the company.

7.3 Converting a tender to a contract

Once the tender has been submitted to the client or their consultants, there is little that the tendering contractors can do except wait. The tender price is usually required to be guaranteed for at least 90 days, but up to 180 days is not uncommon, especially in international contracts where client procedures can be slow. Normal good practice requires that the tenders are evaluated in the minimum possible time to allow appropriate feedback to the contractor if the tender bid is not successful.

In a traditional lump sum contract, it is the tender sum (i.e. the figure at the bottom of Table 7.5 above) which is the important figure. All the associated documents, such as bills of quantities, which build up to that figure are just that – associated documents, at this stage. It is highly likely that the form of tender will only be one page and may state only:

> We, XYZ Construction Company Ltd. Offer to carry out the Works known as xxxx in accordance with the tender documents for the sum of £4,519,232.00.

It is this sum which is being offered by the tenderers for hopeful acceptance by the client, thus creating a formal contract through offer, acceptance and consideration. For further discussion on the legal formation of contract, the reader is encouraged to refer to a textbook on construction contracts.

7.3.1 Tender clarifications and queries

Although a pre-tender clarification meeting may have been held soon after the client's consultants have sent out the tender documents, there may still be areas in the individual contractors' tenders which raise questions in the consultant's evaluation process.

Areas where clarifications will be sought from tendering contractors may include:

a) Arithmetical errors in building up to the tender sum. In this case, the normal practice included in both the NJCC Code of Procedure and the CIOB Code of Estimating Practice is for:
   i) The tenderer to be given details of the errors and afforded the opportunity of confirming or withdrawing the total tender sum. As stated above, it is the total price which is tendered, not the build up to the sum. When the estimator has established the overall financial effect of the arithmetical error, they would be in a position to decide whether to 'live with it' or withdraw the tender offer completely.
   ii) The tenderer may also be given the opportunity of confirming the original offer (with errors in the build-up) or correcting the offer to take account of any genuine errors. Clearly, if the tenderer corrects the offer and this raises the tender price, they will run the risk of losing the tender.

Of course, in both cases above, the tendering contractor will not be aware of the value of the tenders submitted by other companies.

b) Technical and commercial items within the tender submission may not be compliant with the tender documents. In this case, the same principles apply of giving the tenderer an opportunity to confirm that the tender is fully compliant (unless it is specifically stated as an alternative tender) or amending the offer. Many clients, however, have very strict rules on returning to the tendering contractors with such commercial clarifications, in an attempt to prevent the possibility of malpractice.

7.3.2 Tender results

When the full tender evaluation has been carried out by the client's consultants, a recommendation for award will be made to the client's decision-making body, either the tender committee or the responsible senior executive. Normally, the tenderer who is technically acceptable and has provided the lowest price will be chosen, but that will depend on the client's criteria and their standard operating procedures. It is relatively easy to ensure that a submitted tender is the lowest, by submitting a 'suicide bid' – see the two articles below taken from the RICS in-house magazine and *Building* magazine during 2011. As stated in the second article, if a suicide bid is accepted by the client, the contractor will no doubt seek to make up any losses during the construction stage by claims, requests for variations etc., which may not make for harmonious working relationships on site.

Some clients, especially in mainland Europe, have responded to this by always accepting the second lowest bid, clearly hoping there will not be two suicide bids on the same project.

One in four tender responses classed as 'suicide bids' say RICS members

Quantity surveyors have told a RICS survey that they believe more than 20 per cent of the construction tenders submitted during 2010 and 2011 are priced at a sub-economic level.

The survey, conducted by RICS among nearly 400 quantity surveyors, also revealed that most 'suicide bids' are being priced at a point which is around 10 per cent below the true value of the project in question, with some extreme cases seeing costs being pitched as much as 40 per cent below what is considered to be realistic.

The study provides further evidence of how widespread the issue of sub-economic tendering has become, with other findings from the survey showing that:

- 68 per cent believe they've lost contracts because they've gone up against tenders deliberately submitted at a sub-economic level
- 78 per cent think the issue is going to become more of a problem in the next 12 months
- 79 per cent have encountered an increase in 'suicide' bidding in the last three years

- 64 per cent have advised a client not to except a tender because they considered it priced at a level too low to be viable
- 57 per cent have seen their clients accept a bid, even though it was costed at too low a price to be viable
- 44 per cent have encountered tenders from subcontractors that they consider to based on unrealistic costings

In response to both the survey results and the Government's recently published Construction Strategy, David Bucknall – chairman of Rider Levett Bucknall and chairman of the RICS Quantity Surveyor and Construction Professional Group Board – is calling on his QS colleagues to seize the opportunity before them.

He says that the Quantity Surveyors' main responsibility is to point out to the client the risks of sub economic bidding – particularly relevant now as the market is 'bumping along the bottom' with certain key material prices and fuel costs rising sharply.

He explains:

> The QS is being presented with a golden opportunity to prove the value of their professional advice and expertise. If we change and integrate the way we procure construction work, then we can mitigate and in due course eradicate sub economic tendering and still get maximum value-for-money.

Sub economic bidding is the old model; we need to move towards early collaboration and integrated bidding by the whole supply chain. QS's are in a prime position to lead this process by offering their clients accurate information and advice which will help them mitigate risk and operate to the highest of professional standards.

This improved procurement will both tackle the issue of sub economic tendering and start on the road to delivering the 20 per cent reduction in public sector construction costs recorded by the Government Construction Strategy

---

Source – RICS in-house magazine, 20 June 2011

Underbidding: Warning! Highly risky manoeuvre

If you need proof of just how much damage underbidding can do, go to Norwich. This is where the collapse of Connaught last September spelled disaster for a £17.5m housing maintenance contract. When it won the bid, Connaught had been desperately trying to win work to keep its show on the road. Earlier in the year, the bid had been challenged in court by rival bidder Morrison as being 'abnormally low'. Morrison's bid was £5.5m higher.

When Connaught finally fell, the city was rocked. Three hundred jobs were lost, and the council had to draft in replacement firms on emergency contracts, causing disruption to services. 'There were much longer waiting lists for repairs,' says Conservative councillor Nikki George. 'It was the people of Norwich who suffered.'

'Underbidding and trying to claw back costs from the supply chain leads to confrontation, erodes trust, and doesn't work with a partnering culture' – Michael Ankers, Construction Products Association

Of course Norwich was not the only council affected, underbidding was systemic at Connaught and is rife throughout the construction industry. In the end, underbid contracts harm the bidder, the client, the end-users and rival firms that might have been able to do a better job. So why does underbidding continue? And what, if anything, can be done about it?

Almost everyone in the industry agrees that bid after bid is going in at or below the cost of a job, though nobody will admit to doing it themselves. A CIOB survey of its members released in January found that 82 per cent of respondents thought so-called 'suicide bidding' existed in the industry but there is precious little hard evidence on how prevalent the practice really is. Paul Sheffield, chief executive of Kier, estimates that at least 10 per cent of bids are loss-making, while James Wimpenny, regional director of BAM in the North-west, estimates the figure is one in four. Another social housing contractor estimates the figure is more like one in five, and that the undercutting is severe. 'We have seen quite a few examples where people have put contracts in 20 per cent under [the nearest bidder]. In desperate times people do desperate things,' he says.

Consultants, too, are witness to the practice. Richard Steer, senior partner at Gleeds, says: 'In certain circumstances we are seeing very low bids. They can be 20 per cent lower [than the next rival].' He admits underbidding by consultants themselves does go on, but by a smaller margin. Yet the head of one major consultancy privately rages at rivals who offer bids two-thirds lower than his firm's, barely covering the cost of the equipment that was being installed.

Destroying trust

So why do companies underbid? One reason is, like Connaught, to use it as a temporary way to boost revenue and help the firm – albeit temporarily – stay afloat. Wimpenny says: 'People are taking a strategic decision to do it. We know on some bids you just won't get that money back.' Sheffield adds: 'By and large it's about a surge of costs, and to keep people busy.'

But it is not always simply a way to buy revenue in harsh times. Many contractors do it in the expectation of reclaiming costs by exploiting loopholes in the contract and squeezing suppliers. 'I'm not 100 per cent sure that we know we're underbidding. Instead we're just saying that the market is falling and we'll get it back [during the contract],' says Wimpenny. Figures from last year's Constructing Excellence survey show just 47 per cent of projects finished on price, and the figure has not been above 52 per cent during the last decade. The contract seems to be just the start, rather than the end, of price negotiations.

Some in the industry defend low-bidding as a common sense response to a weak economy. Stan Hornagold, the founder of the Marstan Group, argues that it's merely a sign of competition at work if a contractor decides to work for no profit. If supermarket brands sell customers loss-leaders, why not construction firms?

Yet this low-bidding strategy looks increasingly risky in 2011 as tender prices stagnate and the cost of materials and energy (in particular oil) rapidly rise. Michael Ankers, chief executive of the Construction Products Association, says that contractors won't be able to claw back money by squeezing manufacturers. 'You're not going to be able to drive down costs through materials,' he says. Data from the RICS suggests that tender prices will increase by just 0.2 per cent by the end of 2011, well below normal inflation – let alone materials inflation. Copper has risen by 225 per cent since 2009, iron ore doubled in price since last summer, and with Libya now at war, oil is at a two-and-a-half year high.

Source: *Building* magazine, March 2011

7.3.3 Post-tender negotiations

When a contractor has been chosen and approved by the necessary authorities within the client organisation, the client may do one of the following:

a) Accept the tender bid as it stands and forward a letter of award for that amount. This is clearly the cleanest strategy as there is a pure offer and acceptance and therefore a formal contract comes into existence.
b) However, as life is not always that clean, there may be some slight amendments to the scope of works that the client would like to make, and consequently negotiate an appropriate amendment to the contract price.

Much depends on the client's standard operating procedures, whether post-tender negotiations are acceptable, but being able to negotiate with a single firm has many advantages, mainly related to being able to tidy up the scope of works before they are formalised into the contract documents.

In fact, in many procurement routes allied to the PFI, the post-tender negotiation is a recognised stage – named Best and Final Offer (BAFO) when the successful contractor (called the preferred bidder) goes back to the tender build-up to assess if they can provide a more 'competitive' price to the client.

7.3.4 Letters of intent and contract award

Letters of intent are used when the client wishes to appoint a contractor but the final contractual details have not yet been agreed, so the letter of award cannot be sent and contract documents cannot yet be signed. The purpose of a letter of intent is therefore to merely inform the contractor that their tender was successful and that a contract would be entered into in due course. However, modern practice (in the UK at least) has altered the nature of the letter of intent and it may create smaller contracts in its own right. It is relatively easy to see that unless this is carried out by a specialist, both the client and contractor would be well advised to avoid letters of intent. For further discussion, the reader is referred to construction contract law textbooks.

The legal contract is formed when the letter of award is sent from the client to the contractor, or the formal contract documents are signed, whichever comes first.

7.4 Feedback and action

As with any appraisal system, feedback on performance is clearly beneficial in order to try to improve future performance, providing that the feedback is positive, concentrates on areas where there is opportunity for improvement and any personal blame is avoided.

If the tender has been successful and the company is awarded the contract, this does not mean that the tender process was carried out as effectively or efficiently as possible. The tender price may have been significantly lower than the next price without ringing alarm bells within the client organisation. The client or their consultants should always inform all tendering contractors of the number and value of tenders received (with company details anonymised), so that each company can evaluate their own performance in a truly competitive situation. Clearly, if the tendering company's price is significantly lower, it is reasonable to assume that there was some opportunity missed to improve the profit margins, or a mistake may have been made which has not yet materialised. If that is the case, the mistake will certainly materialise during the construction period.

7.4.1 Handover to construction operations

For the successful tenderer, the 'post contract' stage has now begun. The project will be managed from this point by the operational departments of construction management (for technical and logistics issues) and by the commercial/QS department for contractual and commercial issues. There now needs to be a formal handover from the bid manager of all appropriate information on which the estimate and tender were based, so that the operational departments can prepare for mobilisation and construction.

Information required to be handed over would include:

a) any prequalification questionnaires together with the company's responses
b) actual tender documents originally received from the client/consultants, including all amendments and addendums
c) all tender query lists with client responses
d) all correspondence during the pre-contract stage with the client and design team/consultants
e) a copy of the form of tender actually submitted
f) all priced bills of quantities/other pricing documents which built up the tender price; the rates should be adjusted to take account of any negotiations, clarifications or queries during the tender stage
g) calculations for the build-up of rates included in the bill of quantities, including allowances and changes made at tender review
h) pre-tender method statement and pre-tender programme/schedule
i) all quotations received from suppliers and subcontractors, including internal analysis of these quotations
j) project overheads submitted as part of the tender
k) estimator's notes and report submitted to the review committee
l) notes and report from any tender stage site visit
m) any further information received after the tender submission.

In order to avoid confusion, there should be a formal handover meeting and minutes should be taken during this meeting. This is therefore an opportunity for a detailed discussion on decisions taken during the estimating stage, including all technical and commercial assumptions, which could affect construction stage operations.

The project is now handed over to the operational team and the estimating and bid management department can now close its books and reflect on another successful job.

7.4.2 Feedback on tender performance

If, however, the project was not awarded, it is even more important that feedback is received in order to try to improve tendering success rate in the future. As we have mentioned previously, estimating is an expensive business for a construction company and the costs of the estimating and bid management departments form part of the general company overheads. The costs of estimating those unsuccessful projects must be borne by the projects which are successfully awarded, so if the company is successful in one in three projects, the successful projects will carry the costs of estimating three projects on average. However, if the company is only successful in one in eight projects, the successful projects will carry the costs of estimating eight projects on average. This will clearly increase the overhead percentages thereby increasing the tender prices and make the company even less competitive and less likely to win further projects. Poor tendering performance creates a vicious circle.

The NJCC Code of Procedure for Selective Tendering states at section 5.6:

> Once the contract has been let every tenderer should be supplied with a list of the firms who tendered (in alphabetical order) and a list of the tender prices (in ascending order of value)

Most contractors are fully aware of who else is on the tender list, information which is usually provided by domestic subcontractors who may be asked for a quotation from most or all of the tendering main contractors. However, the unsuccessful contractors will be very keen to know how their bid compared with the other tenderers. The Code of Procedure therefore encourages the client to give this information without specifically stating which company submitted which bid, as the company list is alphabetical and the list of tender bids is in ascending order of value. Some major clients do not go this far and merely inform the individual tenderers that their tender price was second lowest, third lowest etc. without giving details of tender prices of the other firms. Information passes around the marketplace remarkably quickly and most companies will find out other companies' tender figures at some point, therefore clients may wish to ensure that this information is as accurate as possible and release the figures whilst still recognising the confidentiality aspect.

In addition, a formal feedback meeting may be arranged, in which the client or their consultants will feed back information on the tender submission, such as:

a) the quality of the tender submission, within any headings given in the tender documents
b) any deficiencies or non-compliances found in the tender submission
c) any qualifications made by the tenderer which were found to be unacceptable
d) any procedural discrepancies found during the tender period

e) recommendations for possible steps which may be taken to improve future tenders.

It is very easy for client organisations to develop an almost arrogant attitude to their dealings with construction contractors, given the overriding and pervading culture of master and servant in the industry. The modern industry is very different from previous generations and the philosophy of a legal contract is that both parties bring a value proposition to the table. In the twenty-first century, the vast majority of clients for major construction works are corporations and they contract with other organisations for services they cannot perform themselves. The client contributes payment (consideration) and the contractor contributes an expertise in building which is required by the client. Working to improve the efficiency of this relationship will benefit both parties in the long run.

7.5 Summary and tutorial questions

7.5.1 Summary

The contractor's estimate is raw data. It is purely an estimate of how much the building is likely to cost in terms of the resources required. This includes direct costs such as labour, materials and plant, together with indirect costs such as preliminaries and general company overheads. The estimator is not responsible for winning work for the company, although depending on the company organisation, the bid manager may have that responsibility.

Once the estimate has been prepared, this is converted into a formal tender by the senior management of the company. The tender is a formal quotation, i.e. a legally binding offer to carry out some construction work as defined in the tender documents (these tender documents are the 'invitation to treat' in legal terms).

In order to convert the estimate to a tender, the company will require a summary of all resource costs plus an estimator's report assessing all the project characteristics. This will then be scrutinised to assess the exposed risks, how much general company overheads should be allocated to the project and what profit margin should be allocated depending on the prevalent economic conditions in the marketplace. Due mainly to the recent economic climate, when the number of available projects has been dramatically reduced, contractors have been tempted to submit tenders at less than actual cost (i.e. at a loss) in the hope that this will have a better chance of winning work and thereby keep their businesses ticking over. This is clearly not a sustainable strategy and carries a great risk of increasing the chance of insolvency. Whilst clients would be equally tempted to accept such bids in the hope of paying less for the project than they originally expected, they should proceed with caution as the costs of dealing with a contractor's insolvency may far outweigh any savings in the tender price.

When all these commercial decisions have been made by senior management, the tender would be submitted on the required date and be evaluated by the client or their consultants.

As only one company can be appointed to the project, the successful company will prepare themselves for delivering the project by handing over all estimating and tender stage information to the operational teams and the unsuccessful companies will seek feedback on their tender performance to help improve future tenders.

7.5.2 Tutorial questions

1. What is the essential difference between an estimate and a tender?
2. What are the major risks which need to be considered for a proposed construction contract?
3. Outline a proposed agenda for a tender settlement meeting.
4. Comment on the advantages and disadvantages of 'suicide bidding'.
5. Discuss the main advantages to a tendering contractor of receiving feedback on their unsuccessful tenders.
6. How would the construction operations departments use the information provided for them by the estimators?

8 The construction stage

8.1 Cash flow forecasting and control

The contractor's income and cash flow payments will depend on the payment mechanisms agreed in the conditions of contract. There are three basic ways that a contractor will be paid for construction work:

a) lump sum (fixed price) contracts, usually based on firm bills of quantities, with payments being made either on completion of milestones or by monthly valuations of completed work
b) re-measurement contracts based on bills of approximate quantities or a schedule of rates
c) cost-plus or cost reimbursable contracts where the contractor will be paid all approved costs plus an element to cover overheads and profit.

The preceding chapters of this book have been based primarily on alternatives (a) and (b) above, as an estimating procedure must be carried out to establish the lump sum and also to establish the rates in a bill of approximate quantities. Therefore, to be able to forecast and control their incoming and outgoing cash flow, the contractor must have a good idea of when they are going to do the work and the appropriate value (incoming cash) and cost (outgoing cash).

This section will therefore assume that the project is based on a lump sum tender with firm bills of quantities, together with a construction programme. In this way, all the relevant information is in place regarding time, value and cost.

The project budgeted cost model has already been set during the estimating and tender stage. The total budgeted income for the project will clearly be the lump sum figure (the 'contract sum' once the formal contract has been entered into), which will be subsequently amended by any authorised variations and changes issued by the client or consultants. The total budgeted costs will be the aggregated resource costs of labour, plant and materials, plus payments to subcontractors. Head office overheads are also a cost to the contractor, but as they are not project-related costs, they will normally be accounted for separately by the company's central finance department. The tender programme will indicate when all the site operations are due to be carried out, thus expending the costs and generating the income.

Taking an example of a simple project with a lump sum of £80,000 and six months duration (Table 8.1), by reference to the construction programme, it is known what operations will be carried out in each of the six months, barring any unforeseen delays etc. By reference to the bills of quantities, a value can be allocated to this work, which is the amount to be paid to the contractor when the work has been properly carried out. This value would represent the first line in Table 8.1.

However, although this amount is properly due to the contractor and represents the earned value to the client (see section 8.3), the contractor will not actually receive this full amount and will not receive anything for probably six weeks to two months after the valuation has been made. Figure 8.1 illustrates the standard payment delays in construction contracts:

a) The contractor carries out work during the month.
b) At the month end, this work is valued by agreement between the client's cost consultant/QS and the contractor.
c) The valuation is presented to the contract administrator, who issues a formal payment certificate which is forwarded to the client.
d) The client pays the contractor within a period stated in the contract. The client may withhold part of the payment but there are strict statutory regulations under UK legislation regarding withholding payments (which equally apply to payments between the main contractor and any subcontractors).

The client will also keep a percentage of the valuation in a 'retention fund' to be released to the contractor in stages on completion of the project.

In some contracts, and in many international projects, the contractor is required to submit an invoice for the approved amount; therefore the payment timetable is slightly different and generally longer, as shown in Figure 8.2.

a) The contractor carries out work during the month.
b) At the month end, this work is valued by agreement between the client's cost consultant/QS and the contractor.
c) The valuation is presented to the contract administrator, who issues a formal payment certificate which is forwarded to the client.
d) The client approves the payment certificate and requests the contractor to issue an invoice.
e) The contractor issues an invoice which is paid by the client under their normal terms of trade, which may be between 30 and 90 days.

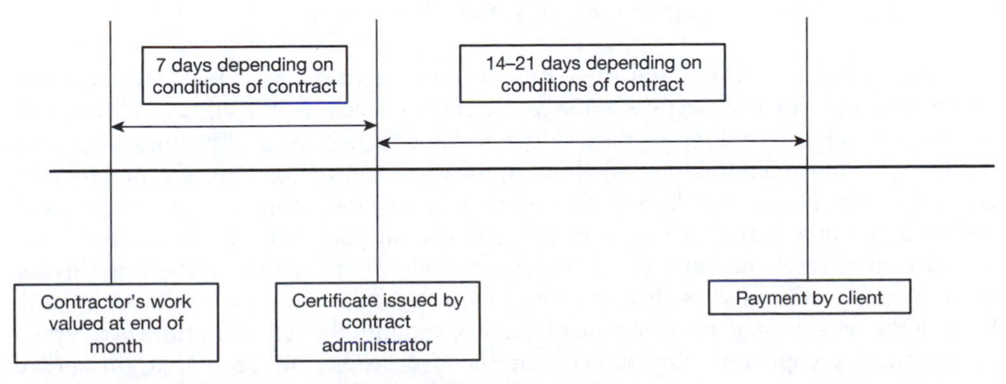

Figure 8.1 Standard payment delays in UK contracts

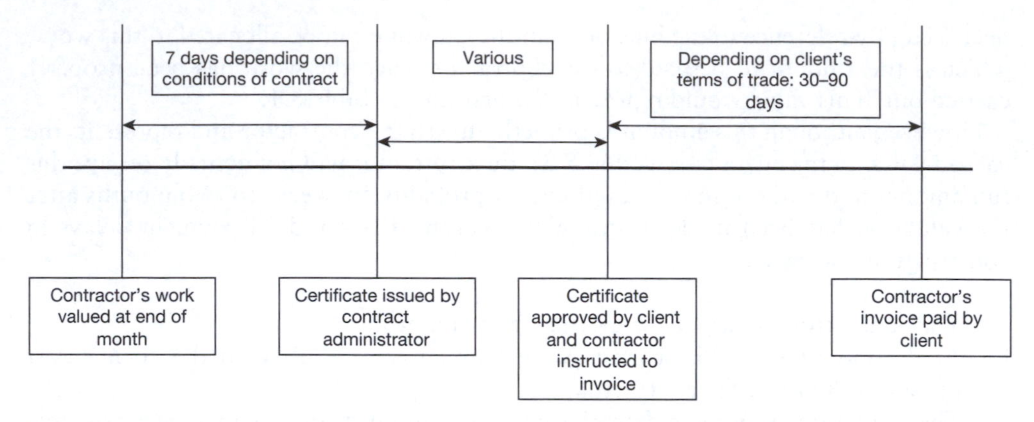

Figure 8.2 Standard payment delays where contractor's invoice required

Returning to the example of the six-month contract, the contractor will have made an assessment of the value of work carried out in each of the six months, recognising that there will be 5 per cent retention on the gross valuations, allowing them to calculate a net valuation in each of the months (Row D in Table 8.1). It is always good practice to show these figures as cumulative, so that any errors made in one month will be self-correcting in the following months.

Given the payments delays explained above, the contractor expects this net valuation of work carried out in month 1 and valued at the end of month 1 to be actually paid in month 3 (Row E in Table 8.1). However, the contractor's costs generally cannot wait this long; as mentioned in Chapter 2, labour and site operatives are usually paid on a weekly basis with subcontractors and suppliers being paid in accordance with the terms of trade of the company, which could be anything between 30 and 90 days. Historically, many contractors have only paid their subcontractors when they have received the appropriate payment from the client (called 'pay-when-paid'). However, this practice is now illegal under UK legislation.

Taking the contractor's lump sum to be broken down as follows:

- overheads and profit: 10 per cent of gross value
- project-related costs paid in month: 30 per cent
- project-related costs paid two months later: 70 per cent.

This would not be unusual and the terms of trade between the main contractor and subcontractors clearly reflect the same terms of trade between the client and the main contractor. Table 8.1 therefore shows the effects of these costs and business decisions on the incoming and outgoing cash flows of the main contractor. Row F represents the cost of work carried out during the month (i.e. less the company's overheads and profit) and rows G and H represent the actual cash paid out by the company, i.e. 30 per cent in the same month and 70 per cent two months later. Calculating this for each of the six months of the project, row J shows that the project is in a negative cash flow all the way through the duration of the project, until the retention fund is returned on practical completion. Remember that the overheads and profit margin spread across all work is a healthy 10 per cent.

Table 8.1 Cash flow budget for a simple project

BUDGETING A CASH FLOW FROM A CRITICAL PATH NETWORK

	1	*2*	*3*	*4*	*5*	*6*	*7*	*8*	\\\\\\\	*13*
A Valuation	22,500	42,500	20,000	165,000	50,000	100,000				
B Cumulative valuation	22,500	65,000	85,000	250,000	300,000	400,000				
C Retention	1,125	3,250	4,250	12,500	15,000	20,000				
D Net valuation	21,375	61,750	80,750	237,500	285,000	380,000		10,000		10,000
E CUMULATIVE CASH DUE			21,375	61,750	80,750	237,500	285,000	390,000		400,000
F Cost of work done	20,250	38,250	18,000	148,500	45,000	90,000				
G Cash paid out	6,075	11,475	5,400	44,550	13,500	27,000				
H			14,175	26,775	12,600	103,950	31,500	63,000		
I Monthly cash paid out	6,075	11,475	19,575	71,325	26,100	130,950	31,500	63,000		
J CUMULATIVE CASH OUT	6,075	17,550	37,125	108,450	134,550	265,500	297,000	360,000		360,000
K MONTHLY CASH FLOW	($6,075)	($17,550)	($15,750)	($46,700)	($53,800)	($28,000)	($12,000)	30,000		40,000

BUDGETING PROFIT FROM A CRITICAL PATH NETWORK

	1	2	3	4	5	6	7	8	\\\\\\\	13
Valuation										
Cumulative valuation										
Cost of work done										
Cumulative cost										
PROFIT										
CUMULATIVE PROFIT										

(Continued)

Table 8.1 (Continued)

BUDGETING CASH FLOW FROM A CRITICAL PATH NETWORK

	1	2	3	4	5	6	7	8	\\\\\\\	13
Valuation										
Cumulative valuation										
Retention										
Net valuation										
CUMULATIVE CASH DUE										
Cost of work done										
Cash paid out										
Monthly cash paid out										
CUMULATIVE CASH OUT										
MONTHLY CASH FLOW										
Interest payments at 0.75% per month										

How can this be? If the project is making a healthy 10 per cent margin spread across all work, why is the company required to subsidise the project until well into the programmed duration? The answer is simple:

PROFIT and CASH are NOT THE SAME

Profit takes no account of when the money is received. The main purpose of cost and financial management in the construction stage is to know how much money is to be received over the course of the project. See Figure 8.3, which shows the standard S-curves for both income and expenditure and makes the same point of costs being paid out quicker than income is received. Hopefully, by the end of the project, the income line will be higher than the cost line, otherwise there are serious problems for the business organisation.

Further salutary points regarding this effect are:

a) The effect gets worse as the S-curve becomes steeper in the mid stage of the project. This is because more work is being carried out in a short time period; therefore the costs going out are higher, which will not be paid for two months.
b) The effect is accentuated in a rising market when more work is being undertaken. Therefore, contractors can and do get into financial difficulties in the good times, just as they do in the hard times.

8.2 Variations and changes

The above analysis has assumed there were no variations or changes to the scope of works or the contract duration (any increase or decrease to the contract period will also have a cost implication). This, however, is not a reasonable assumption as all projects which have ever occurred in the history of the world have been subject to variations and changes, whether they are client imposed, contractor generated (e.g. value engineering proposals), or necessitated through neutral events such as force

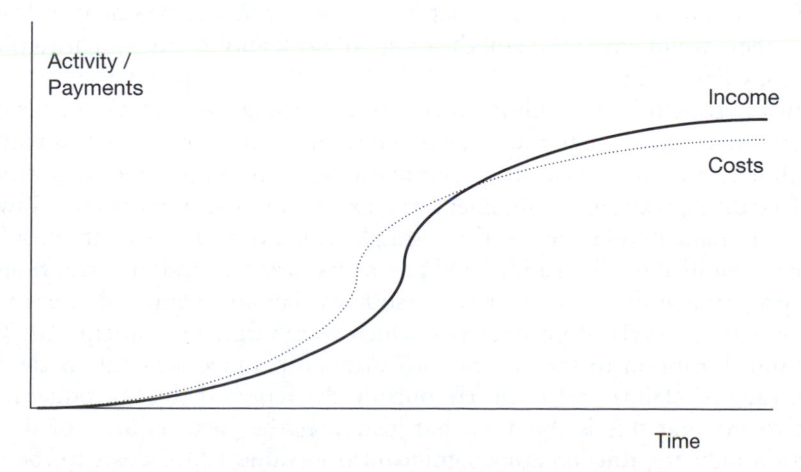

Figure 8.3 Typical S-curves of income and costs on a project

majeure. Most standard forms of construction contracts allow for variations, changes and additions or deletions to be made to the project and include various procedures for the authorisation and valuation of these variations and changes. In fact, legally speaking, if there are no clauses in the contract allowing variations to the scope of work or extensions of time to the completion date, the client would technically be in breach of contract in attempting to vary the scope of works or contract period if the contract is lump sum.

In terms of estimating the cost of variations and changes to the physical scope of work, this can be valued in a number of ways:

a) The change will be measured using the same principles that were used in the original bill of quantities or other pricing document and valued using the rates contained in the document.
b) If there are no direct items and rates which can be applied to the new work, the work may be valued using the same estimating principles and based on similar rates. These are often referred to as 'star rates' as they are often denoted with an asterisk in the final account to show that the rate was not included in the original tender but has been subsequently agreed between the parties.
c) The contractor may be asked to provide a separate quotation for the additional scope of works, which would be evaluated by the client's cost consultants. This is now an increasingly common method of valuation and is given as an option in many standard forms of construction contract.
d) As a last resort, payment may be made using the daywork rates submitted by the contractor in the original bills of quantities. As stated, this should always be a last resort, as payment by daywork effectively puts the new work on a cost-plus basis with a significantly larger overhead recovery by the contractor, when the value of all other work has normally been determined by some form of tendering competition.

Of course, the above only applies to variations and changes which add extra work to the scope. Instructions to omit work which is currently within the scope is relatively straightforward to value, although in certain circumstances and depending on the extent of the omitted work, the contractor may be entitled to payment of loss of the profit that they would have made on the omitted work and to any long lead purchases that they may have made.

In terms of estimating the additional cost of increasing (or decreasing) the contract period without any increase to the scope of work, at a superficial level, this will involve only the time-related charges in the preliminaries section of the bill of quantities, such as hire of facilities, plant and equipment and extra supervision costs etc. However, in practice, it is unlikely ever to be that straightforward and the contractor may be contractually entitled to claim additional payments for delay and/or disruption to their working programme if any of their site resources (labour, plant and equipment) are not working at the levels of productivity which were originally anticipated. The area of delay and disruption to the contractor's original planned schedule is the basis of most contractors' claims and is clearly outside the scope of this introductory book. Suffice it to say that this is the area that generates the vast majority of disputes in construction industry, thus creating significant non-value added costs to the industry (i.e. legal fees).

8.3 Cost value reconciliations

Cost value reconciliation (CVR) is a standard procedure for contractors during the construction stage of a project and is the process of establishing and reporting on the profitability of a construction project. It is important that this is done on a regular basis (usually monthly) so that the senior managers of the construction company have the opportunity to take any corrective measures if they see that the project is not making the anticipated profit margin. By comparing the actual costs with actual revenue at a certain date, the actual profit can be established and compared with expected profit.

Reconciliation is about ensuring that all costs have been accounted for, and all work done has been included in the valuation for payment, in the same way that we would personally reconcile a bank statement with payments we know we have made and income we know we have received. The application of cost value reconciliation in the contracting industry is considered an essential part of cost management for the financial control of construction projects. However despite its undoubted benefits it is often not widely used by either contractors or subcontractors, apart from the larger firms with significant commercial departments. This is generally due to the lack of cost and financial skills necessary to use processes such as these by the company directors who are often predominantly technical.

A well-prepared CVR procedure will show:

- profit and/or loss on each individual contract on a monthly basis
- profit and/or loss for the current accounting period (normally a year)
- current cash position of the project and the company
- details of overhead recovery on individual projects.

This will enable the company to:

- identify any necessary remedial actions and therefore reduce cost liabilities on a monthly basis
- plan future company expenditure more efficiently
- monitor and adjust rates used for estimating based on the profit and loss reporting.

As the whole purpose of a cost and value reconciliation is to match the expenditure with the income, it is important that the following documents related to the project financing are available and kept up to date:

- actual amount of work included in the valuations and certificates issued by the client-side consultants – including an accurate assessment of any over-measures and under-measures, which will give an accurate indication of work carried out to date
- any variations which are not included above and in dispute
- any claims which are pending or in dispute
- materials on site or off site which have been paid for
- subcontractor liability, including any cost build-ups
- any contra-charged invoices
- materials purchase reconciliations
- any cost-to-complete exercise carried out

- estimated final account including estimated profit margin
- all previous architect or engineer payment certificates
- file notes, especially if explaining any of the above.

The contractor would hold regular CVR meetings which would discuss the actual value of work carried out to date less the actual cost of that work, using what is known by accountants as the 'matching principle' (i.e. the costs are matched with the corresponding income). This will give a figure for the profit generated to date. As discussed in section 8.1 above, this income or profit has not necessarily been received yet as it could be awaiting payment by the client; therefore the meeting also needs to be aware of the cash flow position of the project, i.e. how much cash has been received and how much has been paid out. After considering all of these issues, a list of items requiring management action will be developed to ensure the project remains on track.

Any action will generally be required by either senior management of the company or the site team and will concentrate on maximising value, minimising cost and therefore protection of the profit margin. Ensuring timings of payments in accordance with the contract procedures is also essential in maximising cash flow.

8.4 Earned value management (EVM)

Earned value management is basically a similar principle to cost value reconciliation but was developed outside the construction industry and therefore has much wider uses. EVM is a project management technique for objectively measuring project performance and progress during the construction stage and is designed to integrate the physical progress, schedule progress and financial commitment of the works being considered. It basically answers the question 'What did we get for the money we spent?' On a more fundamental level – what bang have we got for our bucks?

The fundamental concepts of EVM are that:

- All project steps 'earn' value as the work is progressively completed.
- The earned value (EV) can then be compared to both actual costs and planned costs in order to determine the current project performance and also to predict future performance trends.
- Physical progress is measured in units of currency (£, $, € etc.), so both schedule performance and cost performance can be analysed in the same terms.

8.4.1 Definitions used in EVM

Like all modern management tools, EVM has surrounded itself with its own jargon, language and terminology, most of which will be reasonably familiar to management accountants as they are broadly similar to the terminology used in the concept of variance analysis. The main terms used in EVM are included in Table 8.2.

8.4.2 Assessing earned value in a project

As with cost value reconciliations, the whole point of the exercise is to try to ensure that the project performs in practice as per the original plan developed at tender stage.

Table 8.2 Terminology and definitions used in EVM

ACWP	Actual Cost of Work Performed	The actual costs incurred in carrying out the work within a given time period.
BAC	Budget at Completion	The total value of the original budget within the contract period. This will be the preliminaries, measured work and any contingencies for provisional work.
BCWP	Budget Cost of Work Performed (**Earned Value**)	The total budgeted value of the work to be carried out in the contract period. This equals the sum of the original budgets for completed work plus the completed portion of provisional work.
BCWS	Budget Cost of Work Scheduled	The sum of the budgets for work scheduled to be carried out in the contract period. May also be referred to as '*planned value*'.
CPI	Cost Performance Index	The cost performance index is the performance ratio comparing BCWP and ACWP, for any given period of time. (BCWP/ACWP)
CV	Cost Variance	The difference between BCWP and ACWP for any given period of time. (BCWP – ACWP)
CWBS	Contract Work Breakdown Structure	The WBS for a specific contract (see WBS below).
EAC	Estimated Cost at Completion	This is the statistical estimate of all costs at completion based upon current performance, calculated by taking actual direct costs allocated to the contract plus a realistic estimate for costs of authorised work remaining. Most common method of calculation is: EAC = ACWP + (BCWP – BAC)/CPI , or BAC/CPI.
EVA	Earned Value Analysis	The principle of measuring and analysing performance by calculating value of a project which has been earned to date. This is the practice of determining how much of the contract budget has been 'earned' on the basis of the actual progress which has been made to date (regardless of the resources deployed to achieve this) and comparing this to the amount of cost incurred and to the planned value.
	Master Schedule	A high-level programme or timetable of events (usually using project milestones), which is agreed between the contractor and the client/consultant for the deliverables within a project. The master schedule is a summary plan which must be traceable to lower-level schedule plans, i.e. at work package and activity level. This can be readily achievable using the level details in programming software such as Primavera.
	Network	A logic flow diagram in a prescribed format consisting of activities and events which must be accomplished to achieve the project objectives. It shows the planned sequence and interrelationships between work in the network. Also referred to as Critical Path Methods (CPM)

(Continued)

Table 8.2 (Continued)

LRE	Latest Revised Estimate	The contractor's latest revised estimate of total costs at completion. This is the sum of the forecast of costs across the contract.
OBS	Organisation Breakdown Structure	A hierarchical structure for the project organisation detailing lines of responsibility through successive levels from senior managers to work package managers.
RAM	Responsibility Assignment Matrix	A matrix formed by the allocation of cost accounts an OBS element, determining responsibilities for all work across the contract.
	Schedule	A timetable of start and finish dates for a list of activities. It is usually calculated from a network plan by a time analysis process, which uses the activity durations in conjunction with any logic constraints.
SOW	Statement (or Scope) of Work	A description of the tasks to be carried out and the deliverables to be produced for a work package or the total project.
SPI	Schedule Performance Index	The schedule performance index is the performance ratio comparing BCWP and BCWS for any given time period. (BCWP/BCWS)
SV	Schedule Variance	The difference between the BCWP and the BCWS for any given period of time (i.e. BCWP less BCWS).
TAB	Total Allocated Budget	Total project budget costs plus client contingencies.
WBS	Work Breakdown Structure	A hierarchical structure which breaks down the work required to develop the main deliverables in the project into successive levels of detail until the lowest level consists of a readily manageable group of tasks, which could be work packages or individual operations.
WP	Work Package	A unit of work at the level where the work is performed and managed. A work package should have a relatively short time period, a budget expressed in money or labour units, a scheduled start and finish. It will, wherever possible, have a set of interim milestones that represent measurable points of physical accomplishment.

By calculating the following ratios, we can get a very good idea about how the project is performing against these original expectations:

a) *Budget at completion (BAC);* which will generally be the tendered lump sum price plus. This is the figure that the client thought the project was going to cost at the start.
b) *Cost variance (CV);* this is calculated as earned value (EV) less actual cost (AC), i.e. the deliverables valued at the original prices less the actual cost. If this figure is positive then project is under budget, which is good.
c) *Cost performance index (CPI);* this is the cost variance shown as a percentage, i.e. EV/AC. CPI greater than 1 is good (under budget); < 1 means that the cost of completing the work is higher than planned (bad); = 1 means that the cost of

completing the work is right on plan (goodish); > 1 means that the cost of completing the work is less than planned (good or sometimes bad). Having a CPI that is very high (in some cases, very high may only be 1.2) could mean that the original plan was too conservative, and thus a very high number may in fact not be good, as the CPI is being measured against an incorrect baseline. As mentioned previously, an overly conservative baseline ties up available funds for other purposes.

d) *Estimate at completion (EAC);* this is the projection of the total cost of the works at the end of the project and can be calculated using the following formula:

$$EAC = AC + \frac{(BAC - EV)}{CPI} = \frac{(BAC)}{CPI}$$

e) *Estimate to complete (ETC);* this is the estimate of costs to complete the remaining works in the project, taken from the date of the estimate. This is calculated as the estimate at completion less the actual cost of work performed to date.

f) *To-complete performance index (TCPI);* this provides a projection of the anticipated performance required to achieve either the budget at completion (BAC) or the estimate at completion (EAC). TCPI indicates the future required cost efficiency needed to achieve a target BAC or EAC. Any significant difference between CPI, the cost performance to date, and the TCPI, the cost performance needed to meet the BAC or the EAC, should be accounted for in the forecast of the final cost.

Clearly, it is no good just calculating these figures and then merely staring at them. A minimum acceptable figure should have been calculated for each factor and set down as a project KPI. It would therefore be the project manager's responsibility to take any necessary corrective action should the factor show an adverse variance from the KPI.

8.5 Cost analysis of completed projects

Once a project has been completed, the cost information in the final account is extremely useful to assist in predicting the costs of future buildings of a similar nature, as discussed in Chapter 5. The cost information thus collected is:

- actual construction costs
- generated in competition
- inclusive of overheads and profit.

The costs, however, do not necessarily represent the actual costs of resources, but they do represent the application of these costs in a competitive environment, which is more realistic and useful for future projects. Notwithstanding this point, cost analyses of a final account are not tender costs, but the costs at completion. Care must be taken when using historical cost analyses about whether they are tender costs or costs at completion (sometimes referred to as 'out-turn' costs).

The Building Cost Information Service of the RICS has designed a set of cost analysis forms for use with completed buildings, which can be downloaded from the public domain section of their website (www.bcis.co.uk) and an example of the summary form is given in Table 8.3. The total cost of the element is given in the first column with the cost per m^2 of gross internal floor area given in the second column. This clearly allows the cost to be transferred to a different building which may have a different

Table 8.3 Summary cost analysis form (source: BCIS website www.bcis.co.uk (c) RICS)

SUMMARY OF ELEMENTAL COSTS					
ELEMENT					
		TOTAL COST	*COST PER m^2*	*UNIT QUANTITY*	*UNIT RATE*
1	**SUBSTRUCTURE**				
2	SUPERSTRUCTURE				
2A	Frame				
2B	Upper floors				
2C	Roof				
2D	Stairs				
2E	External Walls				
2F	External Windows and Doors				
2G	Internal Walls and Partitions				
2H	Internal Doors				
	Total Superstructure				
3	FINISHES				
3A	Wall Finishes				
3B	Floor Finishes				
3C	Ceiling Finishes				
	Total Finishes				
4	**FITTINGS AND FURNISHINGS**				
5	SERVICES				
5A	Sanitary Appliances				
5B	Services Equipment				
5C	Disposal Installations				
5D	Water Installations				
5E	Heat Source				
5F	Space Heating & Air Conditioning				
5G	Ventilation Systems				
5H	Electrical Installations				
5I	Fuel Installations				
5J	Lift & Conveyor Installations				
5K	Fire & Lightning Protection				
5L	Communications & Security				
5M	Special Installations				
5N	Builder's Work in Connection				
5O	Commissioning				
	Total Services				
	BUILDING SUB-TOTAL				
6	EXTERNAL WORKS				
6A	Site Works				

(Continued)

Table 8.3 (Continued)

SUMMARY OF ELEMENTAL COSTS					
ELEMENT					
		TOTAL COST	*COST PER m^2*	*UNIT QUANTITY*	*UNIT RATE*
6B	Drainage				
6C	External Services				
6D	Minor Building Works				
6E	Demolition and Work Outside Site				
	Total External Works				
7	PRELIMINARIES				
8	CONTINGENCIES				
9	DESIGN FEES				
	TOTAL CONTRACT SUM				

floor area but generally similar layout and technology. In order to assess the relative size of the building and elements, the unit quantity and unit rate are given in the remaining columns.

8.6 Summary and tutorial questions

8.6.1 Summary

During the construction stage, the majority of cost management tasks will involve actual costs, because this is the time that the actual costs are being incurred. Rates will also have been agreed and set down in the pricing document at the tender/contract stage, so there is little need for estimating techniques in order to assess the costs of the construction work at this stage.

However, the contractor will wish to continually monitor the costs that have occurred already in the project as well as the reducing balance of anticipated costs for the remainder of the project. Therefore estimating techniques are still valid and appropriate in the construction stage.

The three basic techniques used during the construction stage which require estimating skills are cash flow forecasting and control, cost value reconciliation (CVR) and earned value management (EVM). Pricing and valuing of variations and changes also require an estimating input and finally, at the completion of the project, the actual out-turn costs can be recorded for possible future use by an elemental analysis.

8.6.2 Tutorial questions

1 Outline the essential differences between profit and cash. Why is it important to understand the difference?
2 What would be the effect on a contractor's cash flow if their terms of trade with subcontractors was 30 days given the standard contractual payment terms from the client?

3 Outline the general procedure for valuing of variations to the scope of works of a construction project? Place the procedure in the normal order of preference.
4 Discuss the main purposes of a cost value reconciliation.
5 Who should attend the CVR meeting?
6 Discuss the main purposes of earned value management.
7 Why is earned value normally defined as budgeted cost of work performed (BCWP)?
8 What is the main danger of using cost analysis of a completed project when estimating the cost of a future potential project?

9 Other costs applicable to construction work

9.1 Fees for professional services

Construction is a very fragmented industry and the various roles that are carried out in the design and construction of buildings are often undertaken by different companies from the different professions that contribute to the industry. The definition of a 'profession' can be quite vague and each definition can be slightly different from the others, but as far as the construction industry is concerned, professional services are essentially technical and/or managerial services provided for the project by practitioners who have undergone significant training in their specialist skills and use this skill and training to provide 'solutions' for the client and for the project. The different professional practitioners are also normally regulated by a professional institution that endeavours to ensure the competence and integrity of its members. In all cases, there are very few direct costs of materials or equipment as the direct costs are predominantly skilled labour. All equipment used would be considered as an overhead or indirect cost of the company.

The different professional practitioners which contribute to a construction project are:

- architects
- civil/structural engineers
- mechanical engineers
- electrical engineers
- construction managers
- quantity surveyors/cost consultants
- town planners
- legal practitioners (solicitors etc.).

In addition, local authorities will charge fees for planning permission and building regulations approval, which although not strictly a professional service as defined above, may still represent a significant cost depending on the size of the project.

The costs of these professional services will be borne by the client as part of the total project costs in addition to the costs of the actual construction activity, and clearly must be accounted for by the building owner/client as these people don't come cheap. Under the traditional contractual arrangements, each professional firm will enter into a professional services contract with the client for the provision of their individual services, although the client may also require a 'lead consultant' to manage the services of the other professions.

9.1.1 How professional fees are calculated

Professional fees can be calculated in a number of ways:

a) As a lump sum by assessing the resources required in fulfilling the scope of work of their professional services contract, in the same way as the contractor would calculate the lump sum for the scope of work of the construction contract.
b) As an hourly/daily/weekly rate for each of the grades of professionals engaged on the project, with a percentage addition to cover overheads and profit if this is not already included in the rate.
c) As a fixed percentage of construction costs. This is increasingly uncommon nowadays as the professional fees would automatically increase as the construction costs increase, without necessarily increasing the workload of the professional consultants. Many clients were, not surprisingly, unhappy with this arrangement. Second, clients would enter into a form of bidding procedure by asking consultants to give a percentage deduction from these fixed fees. The firm with the greatest percentage deduction would win the contract.

The way that the fees are calculated in practice will depend mainly on the form of contract between the client and the professional consultant firm. There are several standard forms of professional services contract in use which all calculate the fees slightly differently:

NEC Professional Services Contract (PSC)

This contract would be used if the client chose to use the NEC contract for the main construction works and is therefore designed to fit together with the main contract so that there are as few discrepancies as possible. For a particular project, one main option must be chosen and, like the main NEC contract, there are core clauses and secondary clauses, which when chosen together will provide a complete set of contract conditions.

The main options provide different allocations of risk between the employer (client) and the consultant and also use different arrangements for payment of the fees to the consultant:

- Option A is a lump sum contract in which the risks of being able to provide the services at the agreed prices in the activity schedule are largely borne by the consultant.
- Option C is a target contract in which the financial risks are shared by the employer and the consultant in pre-agreed proportions.
- Option E is a type of cost reimbursable contract in which the financial risk is largely borne by the employer.
- Option G is a term contract in which various items of work are priced or stated to be on a time basis. Therefore the risk of being able to perform the required works at the agreed prices or staff rates is largely borne by the consultant, whilst the client retains control over the individual tasks to be carried out.

The percentage fee type of contract as mentioned above is not included as an option under the PSC as this implies (as discussed above) that the cost of the consultant's services is proportional to the cost of constructing the works. The PSC rejected this option for the following reasons:

- The consultant has no incentive to produce an economical design or other service.
- The cost of construction is largely a function of the market and bears no relation to the cost of professional services.
- The final cost of construction is not established until after construction has been completed, whilst most professional costs are expended much earlier and even before construction starts.
- The effect of variations and changes to the scope on the payments due to the consultant are difficult to assess accurately.

FIDIC Client/Consultant Model Services Agreement (White Book)

This contract has been designed for use with any of the main FIDIC Conditions of Contract and the cost calculations are deliberately left open. The parties are free to agree both the amount of cost and how it will be paid in Appendix C to the contract. This will also cover additional services required from the consultant plus any disbursements and expenses.

Architect fees – The RIBA Agreements 2010

Architect fees usually represent the greatest proportion of all the professional fees on a construction project and will usually range from 3 per cent up to 8 per cent of the construction costs, depending on the extent of design services and whether construction stage supervision is required to be provided.

The RIBA Agreements 2010 is an updated suite of contract documents for the appointment of the architects and are stated to be versatile enough to be used for the appointment of other professional consultants on the project. The suite of contracts has been recognised as an industry standard by attempting to be both fair and balanced to both parties in that it seeks to provide a flexible, user-friendly system for creating professional services contracts. The contracts have been endorsed by the Association of Consultant Architects (ACA), the Royal Incorporation of Architects in Scotland (RIAS), the Royal Society of Architects in Wales (RSAW), the Royal Society of Ulster Architects (RSUA) and the Chartered Institute of Architectural Technologists (CIAT).

The suite of documents includes:

- RIBA Standard Agreement 2010: Architect
- RIBA Standard Agreement 2010: Consultant
- RIBA Concise Agreement 2010: Architect
- RIBA Domestic Project Agreement 2010: Architect
- RIBA Subconsultant Agreement 2010.

In terms of the fee calculations, an optional schedule which can be used with any of the above contracts would be included as an appendix and allows the fees to be calculated in a method agreed by both parties.

Consulting engineer's fees – Association for Consultancy and Engineering (ACE) Agreements

The ACE Agreements can also be used for any consultants providing professional services on a construction project, although they are mainly used by consulting engineers. The agreements have several versions depending on how the professional service is appointed and by whom:

- Agreement 1 – Design edition, where the professional firm provides design services
- Agreement 2 – Advisory, Investigatory and other services
- Agreement 3 – Design and Construct Services
- Agreement 4 – Subconsultancy services
- Agreement 5 – Homeowner
- Agreement 6 – Expert Witness (Sole Practitioner)
- Agreement 7 – Expert Witness (Firm)
- Agreement 8 – Adjudicator.

There are also separate Agreements for use in Scotland as the legal system is slightly different from that in England and Wales.

In terms of fee calculation, Part E of the Agreement allows the fees to be calculated in the following ways:

a) on a time basis (Section E1) in terms of hourly, daily or monthly rates for different categories of staff
b) on a lump sum basis (Section E2), either on the total project or split into stages throughout the project
c) as a percentage of the total project costs (Section E3), again with the possibility of being split into stages
d) as a percentage of the total works costs (Section E4), again with the possibility of being split into stages
e) fees for additional works or disruption (Section E5), which is effectively variations and dayworks for the consultant
f) payment of fees if the consultant arranges the performance by others (Section E6), i.e. a management fee
g) payment of expenses (Section E7) either as a lump sum or at cost with an additional management fee.

This contract therefore gives a thorough coverage of the various methods of payment for professional services.

Quantity surveyor's fees – RICS scale of professional fees

Historically, the quantity surveyor's fees on a construction project would have been calculated using the RICS Scale of Professional Fees, a full copy of which can be found in Spon's pricing books. These fee charges itemised all the normal services carried out by quantity surveyors/cost consultants and gave the fees as a percentage of the construction costs, or a percentage of the relevant estimate if the service was for pre-contract work. The fee scales were formally abolished in December 1998 but no alternative guidelines were ever produced. The main reasons for abolishing the fee scales were:

- Many clients asked for a percentage reduction on the fee scales as a form of competition between service providers. This was seen to be making the profession into a 'cut-price' service provider and not therefore an image which was considered acceptable.
- Having the same scale of charges for all firms providing similar services may be contrary to competition and anti-trust legislation.
- Scale fees may be seen as restrictive, at a time when firms are providing a much wider scope of service to clients.

The recommended fees were split into the following scales:

- Scale 36 – inclusive fees for full quantity surveying services on construction projects. This scale was subcategorised into:
  - Category A – relatively complex works with little repetition.
  - Category B – less complex work with some element of repetition.
  - Category C – simple work with substantial element of repetition.

  Separate fees would be negotiated for MEP work, additional services, time charges etc.
- Scale 37 – itemised fees for quantity surveying services where the client can pick from an itemised 'menu' of services offered. This scale was again split into the following categories:
  a) Contracts based on bills of quantities – pre-contract services
     - fees for production of bills of quantities (varied depending on type of construction)
     - fees for negotiating tenders
     - fees for producing approximate estimates, feasibility studies etc.
  b) Contracts based on bills of quantities – post contract services
     - Alternative 1 – overall scale of charges for post-contract services (varied depending on type of construction)
     - Alternative 2 – scale of charges for separate stages of post contract services; i.e. interim valuations, preparing variation account, cost monitoring service.
  c) Contracts based on bills of approximate quantities – interim certificates and final accounts
  d) Fees for negotiating tenders
  e) Contracts based on schedule of prices – post contract services.
  f) Prime cost contracts – pre-contract and post contract services.

There were additional scale fees for QS services in connection with public sector housing schemes (Scale 40), QS services in connection with housing improvements (Scale 44), QS services in connection with Housing Corporation-funded schemes (Scale 45) and QS services in connection with loss assessment due to fire damage (Scale 46). All of these scales were formally abolished in 1999.

Interestingly, all of these scale charges related to services which were directly part of the 'traditional' procurement route and would have been extremely difficult to relate to the modern industry with its much more flexible operations.

Local authority fees for planning approval and building regulations approval

All construction projects require some form of approval from the competent government authority which covers the location of the project. The authorities will normally have made plans regarding what kind of development is acceptable in certain areas, and will make a charge to developers for giving their approval (or otherwise) of the anticipated development.

Planning permission is normally given firstly as an outline approval, based on schematic designs, and then full approval, once the final requirements are known. The planning authority will charge separately for each of these approvals, which could be a significant cost for a major development. In England, there is a maximum charge of £250,000 for full planning applications. Planning approval is also required for any change of use of existing buildings or when a previous approval has expired. The charges for planning approval are normally paid by the designers and included in the fees charged to the client.

Most countries also have some form of regulation regarding the quality of building construction, which are normally set down in the form of Building Regulations and managed by the same local government authorities. Building Regulations approval is again a two-stage process, with the design being given initial approval and the actual construction on site being inspected to ensure that it conforms to the regulations. The local authority will charge for each stage and it is the responsibility of the builder to pay these charges.

Fees for legal services

The costs of services of lawyers, solicitors, barristers etc. have historically been a contentious area in the construction industry (and most other industries for that matter). At the beginning of the project, there is a considerable amount of work for lawyers to do, in terms of title deeds to the land, negotiating contracts for sale, etc. which is all accepted as part of the value added process of the development. Once the procurement has commenced, all tendering procedures, choice of construction contract and contract administration are best left to the specialist construction firms as any contractual amendments in one area may have unintended consequences elsewhere and the technical experts are better able to manage these consequences. After the construction has commenced, the lawyers' input is usually confined to situations where there is a dispute between the parties, so all costs related to the dispute are non-value added to the project and should be reduced or minimised as much as possible.

In terms of fee calculations, most lawyers charge on an hourly or daily rate for the grade of lawyer working on the issue. Added to these fees will be the lawyer's 'disbursements', i.e. charges and expenses they are required to pay out in the course of their work. It is rare that lawyers will work on a lump sum basis or percentage basis, although they may work on what is known as a 'contingency' basis (i.e. a proportion of any award that their client is made by a court or panel – no win, no fee) for contentious work, where they feel there is a good chance of success.

9.2 Capital allowances

Although this is more appropriate to the client or developer of a project, if a company is incurring capital expenditure on construction work on commercial property, they may be able to benefit from tax relief in the form of capital allowances. These allowances are available to all businesses (the rules state 'a person', including sole traders and partnerships etc., but in reality the vast majority will be companies) who retain property as a fixed asset in their business, including those who occupy their own buildings and those who lease property and adapt it for their own use.

The list of organisations who may benefit from these allowances therefore include:

- developers constructing buildings to hold as investments
- businesses constructing or renovating buildings for their own use
- businesses fitting out leased property for their own use
- organisations making contributions towards costs incurred by others.

Even non-resident landlords (that is, those whose main place of abode is outside the UK) may be able to claim capital allowances, although developers who intend to sell on the property will not be able to directly benefit from the allowances, but the value to a potential purchaser can be calculated and form part of the negotiations with potential purchasers.

The main allowances available on commercial property relate to the 'plant' and 'machinery' installed in the building, but not the land or building itself. Machinery takes its ordinary meaning, but plant is more difficult to identify (in essence, it is business apparatus). Qualifying items would typically include most of the mechanical and electrical services, lifts, escalators, etc. and carpets and other fittings and equipment. The potential list is large and will naturally vary from building to building. Whether or not certain items may be allowable will be influenced by factors such as the use of the building by the business together with the building design. Therefore, the building design can have a major impact on the ability of the client to reduce its tax burden by claiming capital allowances. As the allowances relate to the equipment inside a building rather than the building itself, refurbishment contracts are equally valid as new build contracts.

Capital allowances are given at two basic rates. The 'main' rate at which capital allowances are given is in the form of writing-down allowances at the rate of 18 per cent each year on a reducing balance basis. This is available for assets such as sanitary installations, fire and security alarms, telecommunications and data equipment etc. This means that if these assets were valued at £1 million in a new building, then £820,000 would remain to be written-off by the next year and £672,400 the following year and so on. The capital allowances are based on these valuations. However, a 'special' rate of 8 per cent per cent each year on a reducing balance is available for 'integral features'. These are electrical systems (including lighting); cold water; hot water, heating, ventilating and air conditioning systems; lifts, escalators and moving walkways; and external solar shading. This means that if these assets were valued at £1 million in a new building, then £920,000 would remain to be written-off by the next year and £846,400 the following year and so on. Also, enhanced capital allowances in the form of 100 per cent first-year allowances are available on certain energy-saving and environmentally beneficial (water efficient) plant and machinery. In a small number of designated enterprise zones 100 per cent first-year allowances are available for investment in plant or machinery.

Other allowances that may be available include business premises renovation allowances, and research and development allowances, all of which will depend on the actual use of the building.

The amount of allowances available is based on the actual cost of the plant and machinery included in the construction contract documents, plus any additions for associated builder's work, preliminaries, profit and overheads, and other on-costs such as professional fees. Therefore, it is in the client's or developer's interest to ensure that the appropriate values can be easily extracted from the contract documents, especially the bill of quantities.

9.3 Value Added Tax (VAT) and Construction Industry Scheme (CIS)

9.3.1 VAT

Most construction work carried out in the UK requires VAT to be paid at the standard rate (20 per cent at the time of writing) and this applies to both new work and refurbishment work. If the contractor is registered with Her Majesty's Revenue and Customs department (HMRC), they are required to charge the VAT to the client and would be able to reclaim the VAT paid out to subcontractors and suppliers. However, certain types of work can sometimes be charged at a reduced rate of VAT, or even at the zero rate. The rules for how VAT is charged and at what rate are made by the government of the day and these rules will clearly change over time.

For a VAT-registered construction business, it is obviously important to charge the correct VAT rate and the reduced rate, or zero rate, can only be charged if certain conditions are met. The conditions can relate to different aspects of the work, including:

- the type of building
- the type of work and the equipment installed
- when the work is carried out
- who is the client.

Other aspects of the work can also affect the VAT rate, such as:

- work on ordinary homes
- other buildings which are reduced or zero rated (e.g. owned by charities)
- installation of energy-saving material and grant-funded heating equipment
- installation of mobility aids for elderly people.

VAT is a complex area and the rules are constantly changing, so even the above brief description may well be out of date or not appropriate as you read this section. The various standard forms of contract have significant addendums which relate to the payment of VAT on construction contracts.

9.3.2 Construction Industry Scheme (CIS)

CIS is a withholding tax mechanism which is designed to ensure that those involved in construction are fully tax compliant. The legislation applies to 'construction contracts' (those related to construction operations, but not employment contracts). It obliges subcontractors to register to receive payments from contractors gross or net of 20 per cent tax, or in the absence of a HMRC authorisation and a related payments card, the contractor must retain 30 per cent of the amount payable to their subcontractors engaged to carry out relevant operations.

CIS applies where a 'contractor' (that is, a business that pays subcontractors for construction work) engages a 'subcontractor' to carry out certain services. The term 'contractor' may include property developers, building companies and all associated building trades as well as individuals who are connected with these businesses. It also includes all utility companies and transport companies. A person or company is also deemed to be a contractor where they subcontract all or part of a relevant contract under which they are a subcontractor for CIS purposes. Under the scheme, some businesses, public bodies and other concerns outside the mainstream construction industry but who regularly carry out or commission construction work on their own premises or investment properties are deemed to be contractors if their average annual expenditure on construction operations in the last three years exceeds £1 million.

The scheme covers all construction work carried out in the UK, including: site preparation, alterations, dismantling, construction, repairs, decorating and demolition. A business based outside the UK and undertaking construction work within the UK falls within the scheme.

Services subject to CIS are extensively defined and HMRC practice is to apply a broad interpretation of these definitions and sometimes it may not be obvious whether a particular project includes services which are within the scope of CIS. Services subject to CIS include:

- all construction services including design and build
- EPIC supply and install contracts, e.g. power supply, telecommunications, etc.
- repair, demolition, site preparation and clearance services
- haulage services, crane hire, scaffolding
- agency services related to the provision of labour
- oil and gas exploration, extraction or exploitation.

Whilst CIS is essentially a compliance issue, with numerous forms and returns to be completed, there is significant risk as tax exposures can be substantial where CIS is not operated correctly with the main responsibility lying with the contractor.

References and further reading

Ashworth A (2005); *Cost Studies of Buildings*; Pearson Publishing

Brook M (2008); *Estimating and Tendering for Construction Work* (Fourth Edition); Elsevier Publishing

Buchan R D, Fleming F W E and Grant F E K (2003); *Estimating for Builders & Surveyors*, Second Edition; Butterworth-Heinemann

CIOB (2009); *Code of Estimating Practice*, Seventh Edition; Wiley-Blackwell

Construction Industry Joint Council (CIJC) (2011); *Working Rule Agreement for the Construction Industry*; CIJC

Cooke, B and Williams, P (2009); *Construction Planning, Programming and Control* (Third Edition); Blackwell

Department for Trade and Industry (1998); *Rethinking Building: Report of the Building Task Force* [Egan Report]; HMSO

Flanagan R and Tate B (1997); *Cost Control in Building Design*; Blackwell Science

Greenhalgh B D and Squires G (2011); *Introduction to Building Procurement*; Routledge

Hendrickson C (2000); *Project Management for Construction, Fundamental Concepts for Owners, Engineers, Architects and Builders* (Second edition); Prentice Hall

HMSO (1994); *Constructing the Team, Final Report of the Government/Industry Review of Procurement and Contractual Arrangements in the UK Construction Industry* [Latham Report]; HMSO

HMSO (2007); *The Construction (Design and Management) Regulations 2007*; HMSO

Institution of Civil Engineers (2012); *CESMM4 The Civil Engineering Standard Method of Measurement* (Fourth Edition); Thomas Telford

Jaggar D M and Ross A D (2002); *Building Design Cost Management*; Blackwell Science

Joint Contracts Tribunal (JCT) (2002); *Practice Note 6 Main Contract Tendering*; JCT Limited

Kirkham R (2007); *Ferry and Brandon's Cost Planning of Buildings*; Blackwell

Myers D (2008); *Building Economics: A New Approach*; Taylor & Francis

National Joint Consultative Committee for Building (NJCC) (1995); *Code of Procedure for Selective Tendering*; NJCC

NSR Management; *National Schedule of Rates*; NSRM

Potts, K (2008) *Construction Cost Management – Learning from Case Studies*, Taylor & Francis

Royal Institute of British Architects (RIBA) (2007); *The RIBA Outline Plan of Work*; RIBA

Royal Institution of Chartered Surveyors (RICS) (1988); *Standard Method of Measurement* (Seventh Edition), (SMM7); RICS

Royal Institution of Chartered Surveyors (RICS) (2010); *New Rules of Measurement (NRM)*; RICS

Spon's Architects' and Builders' Price Book 2012; Spon Press

Useful websites

Building magazine: www.building.co.uk
Constructing Excellence: www.constructingexcellence.org.uk
Construction News: www.cnplus.co.uk
National Schedule of Rates: www.nsrm.co.uk
Office of Government commerce (UK): www.ogc.gov.uk
Official Journal of the European Union: www.ojec.com

Index

187097